Dart, R. K.
    Microbiological aspects of pollution
control / R. K. Dart and R. J.
Stretton. Amsterdam ; New York :
Elsevier Scientific Pub. Co. :
distributors for the U.S. and Canada,
Elsevier/North Holland, 1977.
    vi, 215 p. : ill. ; 24 cm.
(Fundamental aspects of pollution
control and environmental science
series ; 2)

Fundamental Aspects of Pollution Control and Environmental Science 2

MICROBIOLOGICAL ASPECTS OF POLLUTION CONTROL

Fundamental Aspects of Pollution Control and Environmental Science

Edited by R.J. WAKEMAN

Department of Chemical Engineering,
University of Exeter (Great Britain)

1
D. PURVES
Trace-Element Contamination of the Environment

2
R.K. DART and R.J. STRETTON
Microbiological Aspects of Pollution Control

Other titles in this series (in preparation):

J.B. OPSCHOOR, H.M.A. JANSEN and D. JAMES
Economic Aspects of Environmental Pollution

D.P. ORMROD
Pollution and Horticulture

R.E. RIPLEY and R.E. REDMANN
Energy Exchange in Ecosystems

W.L. SHORT
Flue Gas Desulfurization

A.A. SIDDIQI and F.L. WORLEY, Jr.
Air Pollution Measurements and Monitoring

D.B. WILSON
Infiltration of Solutes into Groundwater

Fundamental Aspects of Pollution Control and Environmental Science 2

MICROBIOLOGICAL ASPECTS OF POLLUTION CONTROL

R.K. DART and R.J. STRETTON

University of Technology
Loughborough (Great Britain)

ELSEVIER SCIENTIFIC PUBLISHING COMPANY
Amsterdam — Oxford — New York   **1977**

ELSEVIER SCIENTIFIC PUBLISHING COMPANY
335 Jan van Galenstraat
P.O. Box 211, Amsterdam, The Netherlands

Distributors for the United States and Canada:

ELSEVIER NORTH-HOLLAND INC.
52, Vanderbilt Avenue
New York, N.Y. 10017

Library of Congress Cataloging in Publication Data

Dart, R   K
    Microbiological aspects of pollution control.

    (Fundamental aspects of pollution control and
environmental science series ; 2)
    1. Sewage--Purification--Biological treatment.
2. Sanitary microbiology. 3. Biodegradation.
I. Stretton, R. J., joint author. II. Title.
III. Series.
TD755.D37   628.5   77-8685
ISBN 0-444-41589-0

ISBN: 0-444-41611-0 (series)
ISBN: 0-444-41589-0 (vol.2)

Printed in The Netherlands

INTRODUCTION

This book has been written in an attempt to explain the use of micro-organisms in removing pollutants from the environment, and the ways in which they can be utilized to recycle essential materials.

However, micro-organisms can play a role, both advantageous and disadvantageous, to man and his activities. They may be pollutants themselves and cause disease, or they may cause pollution if their presence is encouraged.

In many industrial and engineering processes, it is desirable that engineers have some understanding of micro-organisms, so that full use may be made of their potential and any adverse effects avoided. In some cases the engineer may not immediately recognize the process he has designed because our emphasis has been on the microbiological aspects of the subject and not the processing aspects.

Because many engineers will be unfamiliar with some of the subject areas, we have tried to refer to primary sources of information and to treat each chapter as a self-contained review. This has meant some slight repetition, but as little as possible.

The book covers a wide range of subjects and includes practical aspects such as sewage treatment, and also aspects which at the present moment are mainly theoretical, for example the genetic aspects of bio-degradation of pesticides.

The most recent types of pollution are the production of new materials which cannot be broken down by natural reactions e.g. detergents and chlorinated hydrocarbons. An understanding of the biochemical capabilities of micro-organisms can help in the production of compounds which are both effective and biodegradable e.g. synthetic detergents with biodegradable straight chain alkyl groups instead of non-biodegradable branched chain compounds.

A knowledge of the way in which diseases are spread has led to certain public health and hygiene measures being taken which have doubled the life expectancy in the United Kingdom since 1800. This has, however, meant a loss in other areas e.g. the correct treatment of human excreta required with increased urbanization has resulted in a net loss of calcium and iron from the soil to the rivers and sea.

However, if the biochemical capabilities of micro-organisms are maximized, the waste can often be disposed of by utilization, e.g. paper can be hydrolysed to give sugars, which in turn can be fermented to ethanol by yeasts. Similarly, petroleum waste can be utilized to give yeasts for use as single cell protein for cattle feed. We have tried where appropriate to consider possible new ideas, even if these are only at the level of theoretical consideration.

ACKNOWLEDGEMENTS

We would like to acknowledge the work done by Mrs. Susan Dart who typed the entire book, and also proof read the final version. Our thanks are also due to Mr. John Brennan who did the photography, and prepared some of the diagrams.

R. K. Dart

R. J. Stretton

C O N T E N T S

INTRODUCTION   V

1. MICROBIAL PRODUCTION OF POLLUTANTS   1

2. AIR POLLUTION AND MICRO-ORGANISMS   29

3. HEALTH HAZARDS ARISING FROM WATER-BORNE PATHOGENS   52

4. WATER TESTING   73

5. SEWAGE TREATMENT   112

6. DISINFECTION AND RECYCLING OF WATER   141

7. EUTROPHICATION   150

8. THERMAL POLLUTION   165

9. THE SULPHUR CYCLE AND WASTE RECOVERY   172

10. OIL POLLUTION   180

11. BIODEGRADATION   193

MICROBIAL PRODUCTION OF POLLUTANTS

In the technologically advanced countries water pollution rarely means contamination with potentially pathogenic bacteria, viruses, protozoa or metazoa, but a supply which contains unwanted chemicals. However, micro-organisms may still cause pollution, to a modest or serious level, and this can come from the growth of free living hetero- or auto-trophic micro-organisms which produce undesirable or toxic metabolites. The production of chemical pollutants affects the quality of water, or changes conditions in the soil or atmosphere. Similarly, food for human or animal consumption can become contaminated with toxic microbial metabolites, and made unfit for consumption.

Metabolism of Metals

i) Mercury

The chief raw material for the commercial production of mercury is cinnabar (mercuric sulphide) from which the metal is obtained by heating and condensation, in a state pure enough for most purposes. Mercury occurs naturally in the sea, at levels of 0.1-0.27 µg/l, in the Pacific, the concentration increasing slightly with depth. In ocean waters, mercury is thought to exist as the complex anion, $HgCl_4^{2-}$, and in this form does not appear to collect in bottom deposits, as it does in fresh waters. Lower values of 0.01-0.02 µg/l have been reported for the Solent and English Channel.

There is little data available on unpolluted fresh waters, but the concentration resulting from weathering of rocks is less than 0.1 µg/l, unless there are mercury deposits in the area. River water may contain 0.01-1.4 µg/l, sewage 0.07-2.2 µg/l, sewage effluents 0.2-1.3 µg/l and sewage sludges 0.03-0.75 µg/kg dry weight. The level of mercury detected in fresh water may not be a good indication of the contamination level, because mercury is concentrated in bottom deposits, where it may remain available for many years for uptake by aquatic organisms.

Neurological disorders are evident in dogs, cats and rabbits having a daily intake of 0.4-1.0 mg of mercury/kg of body weight.

Mercury can also enter surface waters in waste discharged from industrial processes, where it is used mainly as the metal in electrical apparatus, and as cathodes in the manufacture of chlorine and caustic soda.

It is also used in the manufacture of vinyl chloride and urethane plastics. Paper and pulp industries use large quantities of phenyl mercuric nitrate which adheres to particles discharged into waterways and is deposited as sediment. The St. Claire River system has received about 20,000 lb. of mercury from industrial activity over a twenty year period [1]. Mercury compounds are also used in paints and as agricultural bio-cides. There may also be contamination of surface waters as a result of atmospheric pollution following the burning of fossil fuels, and the scrubbing of mercury vapour and methyl mercury from the air by rain.

Concern over the environmental behaviour of mercury started following an incident in the Minamata Bay area of Japan. During the period from 1953 to 1960, 116 people were poisoned irreversibly and 43 died from eating fish and shell-fish contaminated with mercury, which came from a vinyl chloride producing factory. There was a similar incident, involving vinyl chloride production in Niigata, where 120 people were poisoned and there were 5 deaths following the consumption of fish. In Sweden during the mid-1960's, mercury discharged in industrial waste either as inorganic compounds, or as phenyl mercury, accumulated as methyl mercury in fish in certain lakes and the resulting high concentrations caused the authorities to prohibit the sale of fish. There was also a decrease in the bird population, in Sweden, which was associated with the use of methyl mercury dicyandiamide as a fungicide.

After the demonstration that mercury in fish was present predomin-antly in the form of methyl mercury, it was shown that unidentified micro-organisms in the natural organic sediment of lakes could methylate mercury. The net result of the process could be mono- or di-methyl mercury, the rate of biological methylation being correlated to the microbiological activity of the sediment. There have been several hypotheses advanced for the mechanism of methylation, but the process is not completely understood. Using a cell-free extract of a methanogenic bacterium, non-enzymatic methylation of mercury has been shown, with methylcobalamin as the methyl donor, in the presence of ATP and a mild reducing agent. The schemes proposed by Wood [2] for methylation of mercury under aerobic and anaerobic conditions are outlined below.

Proposed aerobic methylation of mercury by methyl corrinoid under non-enzymatic conditions.

Possible anaerobic mechanism of mercury methylation in methylcobalamin-acetate synthetase system. (THF = the coenzyme tetrahydrofolic acid)

Mercury can be methylated in a neutral aqueous solution by a purely non-biological reaction, where the methyl donor is methylcobalamin, the reaction being fast and almost quantitative, under both aerobic and anaerobic conditions. However, it has been demonstrated that microbial activity is required for methyl mercury synthesis under natural conditions, unless other methyl metal compounds, e.g. methyl tin, are added. Bacteria isolated from mucous material in fish [3] and Pseudomonas species from soil can methylate mercury under aerobic conditions [4]. Mercury tolerant mutants of Neurospora crassa can methylate mercury under aerobic conditions [5] when an excess of cysteine or homocysteine is present and

4

methylation may be an incorrect synthesis of the amino acid, methionine.
In addition, when mercuric chloride is added to five bacterial species and
three fungal cultures, methyl mercury is produced under aerobic conditions
[6].

The microbial conversion of inorganic mercury compounds into methyl
mercury can take place in the soil under anaerobic conditions, for example,
Clostridium cochlearium has a high capacity for methylation of inorganic
mercury in the presence of cysteine and vitamin B_{12}. The methyl mercury
compounds formed were decomposed by a mercury-resistant Pseudomonas [7].

Although the ability of micro-organisms to methylate mercury has been
shown to exist under both aerobic and anaerobic conditions, the ecological
importance of these observations is difficult to determine, as methylcobalamin is known to be unstable in natural environments. In addition,
trans-methylating activity is inhibited in vitro by cellular proteins and
thiol groups. Anaerobic methylation may not be of ecological significance
because mercury is hardly ever present in nature without hydrogen sulphide
also being present, therefore mercuric sulphide is likely to be formed.
This may explain the absence of methyl mercury in anaerobic mud in certain
experiments [8]. The ability of micro-organisms to demethylate mercury may
also complicate experiments.

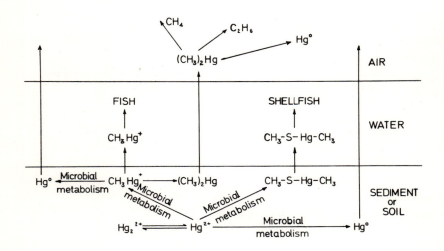

Mercury Cycle after Wood [2]

Anaerobic sediments treated with ionic mercury release elemental mercury. Four strains of bacteria capable of converting the methyl mercury cation to methane have been isolated from lake sediment and E. coli can convert $HgCl_2$ to elemental mercury [9]. Elemental mercury vapour and benzene were products of phenyl mercuric acetate degradation by cultures of mercury resistant Pseudomonas sp. [10].

Hamdy and Noyes [11] showed that a strain of Enterobacter aerogenes was resistant to 1,200 μg Hg^{2+}/ml and could produce methyl mercury from mercuric chloride, but could not produce volatile elemental mercury. The amount of methyl mercury produced was decreased if DL-homocysteine was present in the growth medium. The production of methyl mercury was postulated as a detoxification mechanism.

ii) Arsenic

Arsenic has a long history as a poison, being toxic to humans and animals with a central nervous system, to most higher plants and certain lower organisms. The inorganic ion arsenite (3^+) is more toxic than arsenate (5^+), and the volatile trimethylarsine, $(CH_3)_3As$, is also toxic to humans. There have been cases of poisoning from drinking water containing 0.2 p.p.m. arsenic, whilst in the U.S.A. the recommended maximum for drinking water is 0.01 p.p.m. and the maximum permitted level 0.05 p.p.m.

Arsenite was used for the control of aquatic vegetation and organic arsenicals are still used as herbicides. Lead and calcium arsenates were commonly used as insecticides before 1960. Pattison [12] showed that detergent formulations which contain phosphates may have 70-80 p.p.m. arsenic, and the wash water into which these are introduced may have a level of 0.15 p.p.m. arsenic.

The microbial transformation of arsenic compounds was noticed in human poisoning cases, which occurred in rooms papered with wallpapers using arsenic containing pigments. Growth of fungi resulted in production of volatile trimethylarsine. Challenger [13] showed that Scopulariopsis brevis could produce trimethylarsine from compounds containing trivalent arsenic. From this work a pathway for trimethylarsine production was proposed (see below).

6

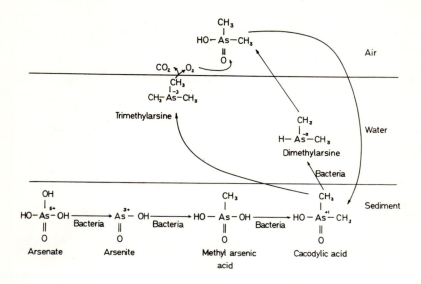

Arsenic Cycle after Wood [2]

Various micro-organisms are capable of synthesising trimethylarsine from
industrial and agricultural arsenic containing sludge [14]. Braman and
Foreback [15] showed that methylated arsenic compounds were found in most
biological materials. The rate of biological formation is probably high
enough to compensate for oxidation of alkyl arsenes to arsenious acid.
However, arsenious acid could in turn be an intermediate in the formation
of methyl arsine compounds.

The organisms capable of carrying out the transformations of arsenic
are not uncommon, for example, Penicillium sp. can generate trimethylarsine
from the pesticides monomethylarsonate and dimethylarsinite. The conversion of arsenate to dimethylarsine can be carried out in soil by
Methanobacterium and possibly by Desulfovibrio [16]. Also arsenate
reduction to the more toxic arsenite can be carried out by Chlorella,
Micrococcus and the yeast Pichia guillermondii [17].

iii) Selenium and Tellurium

Several selenium compounds are toxic, but it is also an essential element for several mammals and possibly man, although there is a small margin of safety. Organic selenium compounds can be converted micro-biologically to inorganic products, whilst photosynthetic purple bacteria will oxidize elemental selenium to selenate.

A biological mechanism for the methylation by fungi was proposed early this century, namely

$$H_2SeO_3 \xrightarrow{H^+} SeO(OH)O^- \xrightarrow{CH_3^+} CH_3SeOH$$

Selenious acid

$$\xrightarrow[\text{+ reduction}]{\text{ionisation}} (CH_3)_2SeO \xrightarrow{\text{reduction}} (CH_3)_2Se$$

Dimethyl selenide

Biosynthesis of volatile dimethylselenide is a major metabolic pathway for detoxifying selenite in animals e.g. the rat.

Tellurium is also toxic, but is not required as a trace element by any animal. Tellurite is the more toxic of the two common anions. Tellurium was found to methylate in the presence of different Penicillium strains and the volatile metabolite was identified as dimethyltelluride [18]. The practical significance in environmental contamination of these observations has yet to be determined, and it is not known whether micro-organisms do modify the element in natural ecosystems.

iv) Lead

Lead can be methylated by micro-organisms to give $(CH_3)_4Pb$ [19], but the ecological importance of this is not established.

v) Transalkylation

Using Pseudomonas, Nelson et al. [20] showed the formation of methyl tin compounds and, in later work, methyl tin and methyl mercury compounds occurred together. The formation of methyl mercury may not be a direct biological methylation, but could be a transalkylation from biologically formed methyl tin compounds [21]. Similarly, methylation of selenium compounds was carried out by Penicillium strains isolated from sewage and added tellurium compounds were also methylated, but only in the presence of selenium [22].

In order for a metal to be of ecological importance in exerting a toxic effect, the metabolized metal must have a tendency for complex formation compared with its original form. The metabolized form should also be soluble in both water and lipid systems, whereas the original compound is usually only soluble in one or the other. The toxic form will also exhibit a changed valency state and volatility compared with the original substrate.

Metabolism of Nitrogen Compounds

i) Ammonia

Robinson and Robbins [23] suggested that the major nitrogenous compound released into the atmosphere is ammonia and almost all of this comes from biological sources, produced mainly by the heterotrophic activity of micro-organisms on land and in the sea. The nitrogen produced as ammonia from biological activity is eight times greater than that released as oxides of nitrogen from all sources combined. Ammonia is not only an atmospheric pollutant, but its production below ground level can adversely affect plant roots.

Atmospheric ammonia can be absorbed by lakes, rivers etc. and so give rise to another pollution problem by enriching surface waters as ammonium or nitrate (following denitrification), and then be used by algae as a nutrient giving a bloom (see chapter on eutrophication). The cost of treating water supplies can increase because of the reduction in the disinfecting power of chlorine by ammonia. Abeliovich and Azov [24] have shown that ammonia at 2.0 mmol at pH 8.0 is toxic to algae in sewage oxidation ponds.

Ammonia is produced during the decomposition of organic material in soil and the microbial hydrolysis of urea. The rate of loss of ammonia is governed by the type of soil, climatic conditions, the presence of vegetation and the application of nitrogenous fertilizers.

Ammonia formation can be appreciable and volatilisation occurs when a field is treated with organic nitrogen compounds which can be broken down microbiologically. This is particularly true in the case of urea (a common fertilizer), as there are many urease containing heterotrophs in soil, and the concurrent rise in alkalinity favours volatilisation. MacRae and Ancajas [25] have shown that if urea is applied directly to the soil surface, there is no time for the ammonia to react with the soil and the loss is very pronounced. This is more marked than if urea is introduced below the surface, when up to 70% of the urea-N can be lost to the atmosphere as ammonia.

There is a high local concentration of nitrogenous material in areas where cattle are intensively reared; the level of ammonia being twenty-fold higher compared with distant sites. The significant part arises from manure undergoing decomposition, and especially urine of which 90% may be converted to ammonia and volatilised during a week [26]. Hutchinson and Viets [27] showed that sufficient ammonia was absorbed in a lake 2 km away from a large cattle rearing area, to raise the level above that required for algal bloom formation.

The gas can also be evolved during the decomposition of plant remains in soil, and the breakdown of sewage and other organic material in water during the reduction of nitrate.

ii) Nitrate

Nitrate is the end product of microbial breakdown of organic nitrogen in aerated environments in soil and water. The organic substrates are attacked and the nitrogen is released as ammonium salts. Where oxygen is present and the pH is not too low, the nitrifiers oxidize the ammonium ion to nitrate. This may be leached out by percolating water and could be carried to wells or to surface waters which ultimately may be used for drinking. Because of the increased use of synthetic fertilizers in agriculture, combined with the growth of urban areas and the associated development of industrial centres, there has been an increase in the amount of nitrogenous material for microbial attack concentrated in a smaller area. The levels of nitrate in drinking water have risen to approach toxic levels in several areas. There was concern in East Anglia during the very hot and dry summer of 1976 that the level in drinking water could reach danger level for infants.

If the level of nitrate in water rises above 22 p.p.m. [28] there is a risk of methaemoglobinaemia, a disease of infants (particularly those less than six months old) and livestock. A total of 2,000 human cases, many of which were fatal, have been linked with drinking water polluted with nitrate [29]. The World Health Organization have recommended that water for human consumption should contain no more than 10 p.p.m. nitrate-N. In Illinois, where water has been sampled regularly since 1945, the streams contain in excess of 0.3 p.p.m. nitrate-N and in the reservoirs which they feed, the concentration may exceed 10 p.p.m. The area concerned is one in which fertilizers have been widely used [30].

A second major cause for concern is eutrophication, the enrichment of natural waters with nutrients. Another hazard which has been postulated following the consumption of nitrate-rich water or food, is that associated with the in vivo synthesis of nitrosamines (see below).

Nitrate pollution can occur in the intensive rearing of cattle and poultry. It has been estimated that 30,000 cattle will give a nitrogenous waste equivalent to a city of 250,000 people, and the droppings in a large poultry farm can be equivalent to 25,000 people. Few, if any, economical methods exist to treat this amount of waste. Webber [31] has shown that in such circumstances the nitrate load may go up to 50 p.p.m. The application of sewage sludge to the land can also lead to nitrate pollution [32]. This nitrate may move with percolating groundwater and contaminate drinking water supplies.

iii) Nitrite

Nitrite is an intermediate in autotrophic nitrification and in the reduction of nitrate to ammonium for assimilatory purposes. Nitrite does not frequently accumulate because its rate of formation is usually less than the rate of metabolism, although under certain circumstances this may not happen and it can accumulate.

Nitrite is the toxic species in methaemoglobinaemia and it is produced from nitrate by the gut microflora. Selenka [33] has examined the role of bacteria in the formation of nitrite in canned baby foods.

The presence of nitrite is also inhibitory to the growth of higher plants and Curtis [34] showed that avocado and citrus plants are affected by 5.0 p.p.m. nitrite at pH 5.0; 10 p.p.m. at pH 5.5; and 20 p.p.m. at pH 6.0.

Nitrites, and the resulting nitric oxide (NO), are important as additives in meat curing. Under acidic conditions, nitrite forms nitrous acid which can undergo decomposition with the formation of nitrogen dioxide (NO_2) and nitric oxide. The nitric oxide reacts with the myoglobin of meat to form the precursor of myochromogen which is the red pigment in cured meats. In hydrated environments the nitrogen dioxide will give nitrate and nitrite [35].

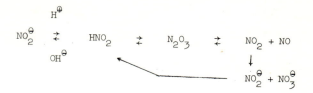

iv) <u>Hydroxylamine</u>

The potent mutagen hydroxylamine may be present in bodies of natural water, as Tanaka [36] observed that it could be detected in the water of a lake in central Japan, at times of the year when oxygen was deficient in the lower regions of the lake.

A mechanism for hydroxylamine biosynthesis has been observed in axenic cultures, and in samples of natural systems it may be a phase of hetero-trophic nitrification. The observations were made with a sewage-derived strain of <u>Arthrobacter</u> which oxidised ammonium to hydroxylamine under normal growth conditions and at several pH values. Hydroxylamine-N reached 15 p.p.m. in growing cultures, however, resting cells of <u>Arthrobacter</u> produced up to 60 p.p.m. hydroxylamine-N when they were provided with ammonium. They would also produce hydroxylamine from acetamide, glutamine or glutamate, but not from glycine or l-aminoethanol. The process requires a source of organic carbon, and the ammonium may have to be bound to an organic compound before it is oxidized to the hydrox-amate which is then broken down, by the organism, to give hydroxylamine [37,38]. Similar results have been obtained with sewage, river or lake water when supplemented with acetate or succinate [39].

v) <u>Nitrogen Oxides</u>

Intoxications resulting from nitrogen oxides, produced biologically, have been reported among farmers making silage from nitrate rich plants. During an early phase of the fermentation of the tissues of plants, e.g. oats, some of the excess nitrate is converted to nitric oxide (NO) which on exposure to air is oxidized to toxic nitrogen dioxide (NO_2), and poisonings and death have followed [40,41]. Wang and Burris [42] demon-strated that the level of nitric oxide in a silo filled with maize may reach 9% v/v and could reach ca. 47.2% v/v in a silo which contains maize tissues such as cobs and husks. The enzymes in the plants, however, may be involved as well as bacterial action.

vi) Nitrosamines

These compounds are known to be carcinogenic, teratogenic and mutagenic at the level of approximately 1 p.p.m. Many of the widely used pesticides are amines which could be nitrosated or microbiologically con verted to products which are able to undergo such a reaction. Nitrilo-acetic acid (a tertiary amine), which in some countries has been included in detergent formulations, could be discharged into natural bodies of water reaching very large tonnages, and would then be available for microbial transformation.

Nitrosamines can be generated in the human gut, provided that either nitrite or nitrate are available. Alam et al. [43] have shown that if nitrate and piperidine occur together in gastro-intestinal tract contents, they give rise to nitrosopiperidine. Similarly, Braunberg and Dailey [44] showed proline and nitrite were converted to N-nitrosoproline by rat intestinal contents. Also, when rats were fed with aminopyrine or heptamethyleneimine and nitrite they developed malignant tumours. The work is not conclusive in showing that micro-organisms are the nitrosating agents, because nitrosamines will be formed from nitrite under low pH, or neutral or sometimes alkaline pH conditions. Some of the species of micro-organisms may be important for the creation of favourable acid conditions for nitrosation.

Hawksworth et al. [45] showed that five out of ten strains of E. coli formed nitrosamines from diphenylamine, dimethylamine, piperidine etc. in a nitrate containing medium at pH 6.5 and above. Similarly, 10% of Clostridium strains, 12% Bacteriodes, 40% of enterococci and 18% of isolates of bifidobacteria will nitrosate diphenylamine.

Ayanaba et al. [46] showed that cell-free extracts of Pseudomonas an Cryptococcus can form N-nitrosodiphenylamine from diphenylamine and nitrite. Mills et al. [47] showed that only actively growing Pseudomonas stutzeri out of 38 cultures tested formed dimethylnitrosamine from dimethylamine and nitrite, and boiled cells of this organism also had this property. The reaction could be carried out by resting cells of other organisms and a cell-free extract of E. coli.

If nitrosamines are formed in soil this could be important if the water travels and is used for drinking purposes or for raising fish.

The problem of nitrate in drinking water is important in this context, because the number of deaths from stomach cancer in a town in England, where the water supply contains nitrate in excess of that deemed safe by most public health standards, is 32% greater for males and 62% greater for females than in towns where the water supply is unpolluted with nitrate [48]. This may be chance or may be nitrosamine formation. The relationship between nitrates, nitrosamines and gastric cancer has been studied in Columbia [49] and in Illinois [50].

Metabolism of Sulphur Compounds

i) Hydrogen Sulphide

Eriksson [51] suggested that the decomposition of organic materials on land yielded 11.2×10^7 tons of hydrogen sulphide to the atmosphere each year and Marchesani et al. [52] proposed that 0.07 tons of hydrogen sulphide per 1,000 square miles was emitted each day from natural sources in the U.S.A. Grey et al. [53] suggested that the most important source of atmospheric sulphur, after industrial discharges in the vicinity of Salt Lake City is the microflora of the lake, the riverbottom mud and marshes near Great Salt Lake.

Hydrogen sulphide can be formed in two ways by micro-organisms; by the reduction of sulphate by Desulfovibrio and related bacteria, or by the breakdown of organic molecules. The sulphate reducers, Desulfovibrio, are ubiquitous in mud, swamps and poorly drained soil. They are unable to utilize any substance other than sulphate for respiration and they show little activity when dissolved oxygen is present or the pH is low. Many micro-organisms can break down organic compounds and release hydrogen sulphide, this being one of the tests used in the identification of micro-organisms. The reaction, which is found in aerobic and anaerobic, psychrophilic and thermophilic bacteria, actinomycetes and fungi, can be represented:

$$HS\text{-}CH_2\text{-}\underset{\underset{COOH}{|}}{\overset{\overset{NH_2}{|}}{CH}} + H_2O \xrightarrow[\text{desulphydrase}]{\text{cysteine}} CH_3\text{-}\underset{\underset{COOH}{|}}{C}\text{=}O + H_2S + NH_3$$

cysteine   pyruvic acid

Sulphide is generally formed slowly in the soil, but the rate increases significantly if organic material is added. A great deal of sulphide produced is precipitated as iron sulphide in the soil and so will only affect plants or animals in soils which are deficient in iron or other cations which will precipitate as the sulphide. The rate of sulphide production is increased if sulphate is present and the environment becomes anaerobic.

Hydrogen sulphide is toxic at the same level as hydrogen cyanide, and 0.1 p.p.m. in water will affect newly hatched fish fry, fry growth and the survival of fish eggs. This is often seen, as the fish fry and eggs are localised at the mud-water interface [54]. Similarly, the reduction in the level of nematodes, in flooded environments, can be correlated with the hydrogen sulphide level [55]. Higher plants are also sensitive to hydrogen sulphide, for example, citrus roots are susceptible to 2-3 p.p.m.

In addition, hydrogen sulphide is offensive, the odour threshold is 0.13-1.0 p.p.m., and contributes to the odours from heaps of animal waste, and near sewage plants. It caused a nuisance in resort hotels up to 1,700 metres away from a faulty sewage stabilization pond, the air level being 0.36 p.p.m. [56]. Also, it caused a nuisance in a town in England, where anaerobic conditions developed in a quarry of an abandoned brick works. The hole filled with water in the gypsum clay. The level of hydrogen sulphide was detectable 1,000 metres away, and the nuisance persisted until the area was filled with power station fly-ash (unpublished observations).

Apart from the direct attack of sulphate on concrete, damage to sewers may be caused indirectly by the hydrogen sulphide produced by sulphate reducing bacteria, or by breakdown of organic sulphur compounds. In this form of attack the corrosion occurs only above the level of the sewage, where sulphuric acid is formed from the hydrogen sulphide. This only occurs when the sewage is anaerobic, which may be the case if the gradient of the sewer is so flat that the sewage receives little aeration and deposits of organic sludge accumulate. Quite small traces of hydrogen sulphide developed in this way may be liberated at a point where considerable disturbance of the liquid occurs, e.g. at a backdrop, and corrosion may then occur in the vicinity.

ii) Sulphur Dioxide and Carbonyl Sulphide

Micro-organisms are capable of producing volatile sulphur compounds other than hydrogen sulphide. The evolution of sulphur dioxide has been demonstrated in vitro and it may be liberated in vivo. In soil samples, cysteine can be converted to sulphate by a pathway involving the formation of cystine disulphoxide and cysteine sulphinic acid [57], and cysteine sulphinic acid may be the key intermediate in sulphur dioxide formation viz.:

$$HOOS-CH_2-\underset{\underset{COOH}{|}}{\overset{\overset{NH_2}{|}}{CH}} \longrightarrow SO_2 + CH_3-\underset{\underset{COOH}{|}}{\overset{\overset{NH_2}{|}}{CH}}$$

The enzymatic breakdown giving sulphur dioxide can be carried out by Neurospora crassa, E. coli and Algicaligenes faecalis [58,59].

Carbonyl sulphide (S=C=O), which is toxic to fungi and to the central nervous system of mammals, is produced during the anaerobic decomposition of cattle manure by bacteria [60]. It is also formed in the soil from the pesticide disodium ethylenebisdithiocarbamate (nabam) [61].

iii) Organic Sulphur Compounds

Methyl mercaptan (CH$_3$SH) which can also be synthesized biologically has a very powerful offensive odour, is phytotoxic and can be produced from methionine by Achromobacter starkeyi, Pseudomonas, actinomycetes and yeasts [62,63,64].

Francis et al. [65] snowed that soils under anaerobic conditions in vitro give off dimethyl sulphide and dimethyl disulphide. White et al. [66] showed that a variety of compounds including methyl mercaptan, dimethyl sulphide and diethyl sulphide, are produced from dairy manure.

Production of Simple Carbon Compounds
i) Ethylene

Ethylene is important as an atmospheric pollutant because it can regulate several physiological events in higher plants. At low concentrations it can cause abscission of leaves, flowers and fruits, enhance the degradation of chlorophyll, increase respiration rate, cause flower droop, inhibit the growth of leaves and shoots, and accelerate the ripening of fruits. Under the ground it can influence root elongation and lateral root formation and may inhibit the growth of soil fungi.

Under anaerobic conditions, soils will produce quantities from 0.6-24 µg/kg soil in ten days. This process is largely eliminated by sterilization, suggesting the involvement of micro-organisms. Ethane, propane and propylene can be produced under similar conditions.

The evolution of ethylene can be inhibited by oxygen and nitrate. The concentration of ethylene in soil varies with the season of the year, but can often be high enough to slow down the development of plant roots. The production can be increased by the addition of certain organic compounds [67].

ii) Carbon Monoxide

Although most of the carbon monoxide is derived from combustion processes, photosynthetic and heterotrophic micro-organisms in the sea or on land can contribute to its production. Weinstock et al. [68] calculated that the annual rate of carbon monoxide production is 5×10^{12} kg which is greater than the 2×10^{11} kg calculated by Robbins et al. [69] as produced by combustion. So the production from natural sources may be significant.

Robinson [70] observed that carbon monoxide was produced in significant amounts from soil incubated in the laboratory. More recently Dobrovol'skiy et al. [71] found small, but easily detectable levels of carbon monoxide, in the soil atmosphere.

Analysis of surface water samples from the Atlantic and Pacific Oceans showed that they were supersaturated with respect to carbon monoxide in the air, so transfer from ocean to atmosphere must take place. The production of carbon monoxide is related to the presence of light as Lamontague et al. [72] showed that more was produced in daylight and the production followed a diurnal cycle. It may be derived from metabolic activity of marine micro-organisms or it could be an entirely non-enzymatic photochemical reaction [73].

iii) Organic Acids

Certain phytotoxic compounds e.g. p-coumaric, ferulic, vanillic, p-hydroxybenzoic, syringic and protocatechuic acids, which may reach 5 p.p.m. in soil, can be produced by microbial decomposition processes. They may also inhibit seed germination and growth [74].

Production of Mycotoxins

The discovery of aflatoxin in 1961 inspired a large amount of research into the chemistry, biochemistry, mycology and food technology of this and other mycotoxins. The contamination of foodstuffs by moulds is a constantly recurring phenomenon, but attitudes to this have changed from time to time. For a long time moulds were regarded as harmless, particularly following the discovery of penicillin and other mould metabolites with powerful curative properties, and no-one considered that similar metabolites could be toxic to higher organisms. Mild fungal contamination of food was considered a regrettable nuisance which detracted from the appearance of the product, and perhaps caused some loss or spoilage; in severe cases the product might be classed as unfit for human consumption and used for animal feed. On the other hand, fungal cultures have been used traditionally to produce some fermented foods.

Work had been carried out on the effects of mycotoxins before the aflatoxin era. However, these were concerned mainly with veterinary problems or confined to one geographical area. There was no consideration of the role that mycotoxins could play in human or animal nutrition and disease. The ingestion of mould toxins does not usually produce dramatic symptoms, but there may be a chronic, vague ill-health. It is often difficult to connect this with the consumption at an earlier date of food contaminated with mould.

There are a few general principles which characterize mycotoxic situations [75]:-

(a) they frequently arise as veterinary problems, and the true cause may not be immediately identified;

(b) the disorders are not transmissible from one animal to another, i.e. neither infectious nor contagious;

(c) treatment with drugs or antibiotics has little effect on the course of the disease;

(d) the outbreaks are usually seasonal, as particular climatic sequences may favour toxin production;

(e) careful study indicates specific association with a particular foodstuff;

(f) examination reveals signs of fungal activity on the foodstuff.

One mycotoxicosis, ergotism, has been known for a long time. Epidemics were recorded in the fifteenth century and fungal growth was implicated in the malady before 1800. The ergot alkaloids (whose structures are shown below) are the causative agents.

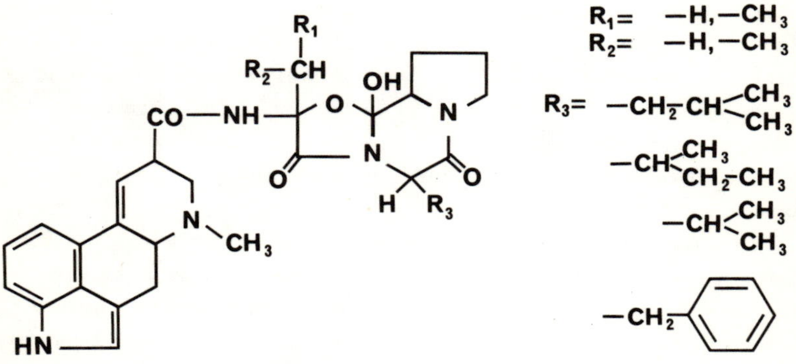

The Ergot Alkaloids (ex Claviceps purpurea)

The active constituents of ergot (the sclerotium of the fungus) are alkaloids based upon lysergic acid and are produced by Claviceps purpurea growing on rye, although the related species C. microcephala, C. nigricans and C. perspali also occur and most cereal grasses may be affected under certain conditions [76,77]. Acute poisoning with ergot is rare, a single dose, except in the pregnant, causes only slight effects. The continued ingestion of small quantities in bread made from infected grain will produce the classical symptoms of ergotism.

In pregnant females the alkaloids cause premature labour followed by a miscarriage.

Toxicity associated with 'yellowed' rice was the subject of investigation in Japan for some twenty years, and after the Second World War the cause was traced to several toxin producing species of Penicillia. Later, two compounds islanditoxin and leuteoskyrin were isolated from culture media.

Islanditoxin

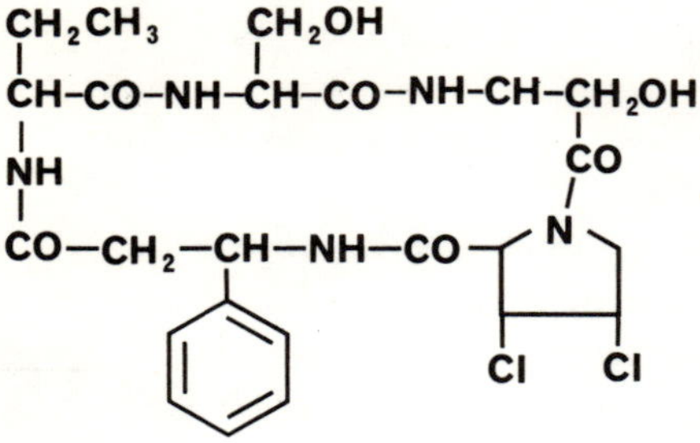

Luteoskyrin

CH₂CH₃   CH₂OH
CH-CO-NH-CH-CO-NH-CH-CH₂OH
NH   CO
CO-CH₂-CH-NH-CO   N
                          Cl   Cl

ex *Penicillium islandicum*

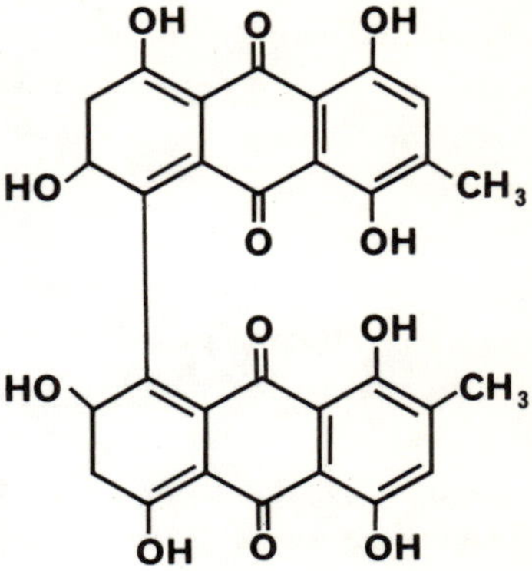

Islanditoxin is a very powerful liver poison which can cause death within 2-3 h, with severe liver damage and haemorrhage. Leuteoskyrin is also a liver toxin, but is slower acting and less potent than islanditoxin.

Other mould species found on rice also produce other, different, toxic compounds, e.g. *Penicillium toxicarium* and *P. citreoviride* produce citreoviridin, a yellow fluorescent compound which is neurotoxic, causing acute paralysis and respiratory failure.

Citrinin

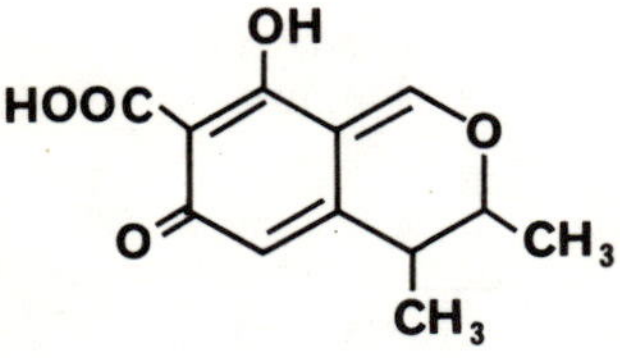

Citreomycetin

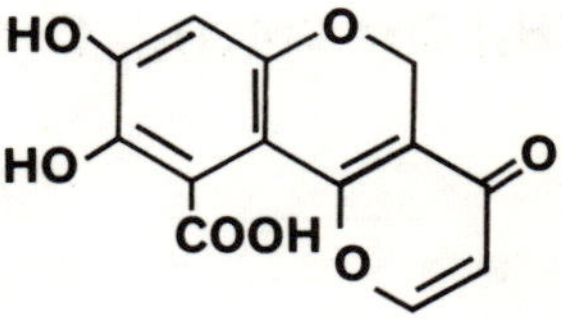

Citreoviridin

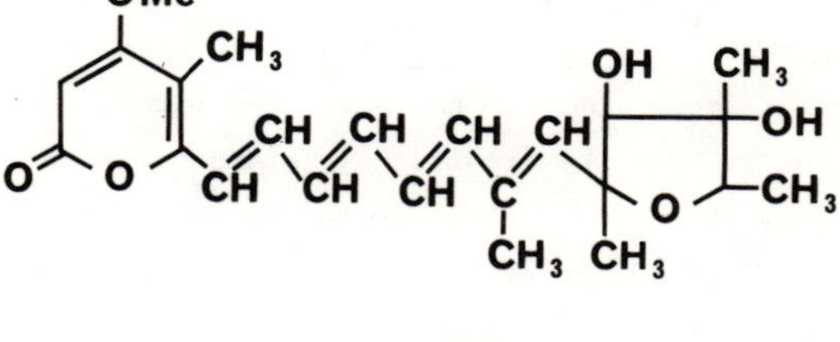

ex *Penicillium citreoviride*

Citrinin produced by <u>Penicillium citrinum</u> and a number of other species, acts predominantly on the kidneys, interfering with water reabsorption, and the related citreomycetin, also produced by <u>Penicillium</u> species and some <u>Citreomyces</u> species, is unusual in being cumulatively rather than acutely toxic, causing chronic kidney damage on prolonged ingestion.

A very different mycotoxicosis is alimentary toxic aleukia (ATA) [78] which is characterized by grave changes in the blood, including leukopenia, agranulocytosis and exhaustion of the bone marrow. The products of <u>Fusaria</u> species, particularly <u>Sporotrichoides</u>, are responsible for the disease. The fungi infect cereals, particularly millet, that have been allowed to over-winter on the ground, e.g. in Siberia. Suitable conditions for toxin formation may arise in the spring when temperatures are in the $1-4^\circ C$ range. The compounds involved in this mycosis, which is without tropical connections, are fusariogenin glycoside and the cladosporic acids.

ex <u>Fusaria</u> sp.

Fusariogenin glycoside

$(C_5H_9O_4)(C_6H_{12}O_5)_2-O$

$CH_3(CH_2)_9CH=CH(CH_2)_n C-SH$

Epicladosporic acid   n = 13
Fagicladosporic acid   n = 9

Strains of <u>Aspergillus</u>, <u>Penicillium</u> and <u>Mucor</u> proliferating on sweet clover may synthesize 3,3'-methylenebis(4-hydroxycoumarin), a compound which is responsible for a fatal haemorrhagic disorder in cattle and sheep eating the mouldy plants [79].

The common mould, <u>Aspergillus ochraceus</u>, invades stored wheat when its moisture content exceeds 16%. It has also been reported as a constituent of the flora used to make the traditional Japanese fish product, katsuobushi. Workers in South Africa obtained ochratoxin from this organism, the compound is highly toxic, having an LD_{50} in the duckling of 0.5 mg/kg body weight.

Patulin
ex Penicillium patulum

Ochratoxin A
ex Aspergillus ochraceus

When the soil is wet and crop residues remain, Penicillium urticae will proliferate and, under certain conditions, patulin appears. This potent phytotoxin does appreciable harm to nearby plants. Penicillium urticae synthesises this compound in culture, and soils deliberately inoculated with the fungus synthesise patulin which is deleterious to test plants [80]. Patulin formation may be implicated as a problem in old apple orchards, or in nurseries where apple trees were previously grown. New trees had shortened tap roots and poor growth, and soil extracts gave the same effect.

The general history of the discovery of aflatoxin and its properties has been reviewed by Austwick [81]. The disorders observed in turkeys, ducklings, pigs and calves were eventually traced to the use of particular batches of peanut meal used in mixed feeds; subsequent work showed that this meal was contaminated with a toxic material which could be traced to the growth of Aspergillus flavus on the nuts from which the meal was produced. Aflatoxin is not a single compound but a group of closely related A. flavus metabolites. At least eight compounds are generally recognized.

For most practical purposes the term 'aflatoxin' is taken to mean aflatoxin B; this is justifiable because it is usually the most abundant and is the most toxic of the group.

In most cases, mould development producing the toxin takes place in the immediate post-harvest stage, during the few days that elapse between lifting the nuts from the ground and drying down to a moisture level below 8%, at which value, fungal growth no longer occurs. Under certain circumstances, the mould attack can occur in the ground before lifting, after damage by insects, or at a later stage when the nuts are in storage or in transit, because of damp conditions. The suppression of fungal activity by drying does nothing to reduce the toxin content following growth of A. flavus on the nuts.

The Aflatoxins (ex Aspergillus flavus)

The occurrence of aflatoxin in peanuts affects their use for direct edible, or for manufacturing purposes.

The nuts used for edible purposes are usually carefully controlled at the growing stage, defective kernels are eliminated, and the toxin levels in consignments of nuts or processed products e.g. peanut butter, is checked. Dickens and Whitaker [82] discussed the merits of an electronic colour sorter for detection of aflatoxic nuts, and concluded that careful hand sorting was more reliable, although in certain instances the sorter was useful.

When aflatoxic nuts are crushed only a small proportion of the toxin comes out in the expressed oil, the residual cake retaining the main fraction. In commercial refining the crude oil is treated with a hot alkali wash and bleaching earth, which effectively removes any residual aflatoxin. This situation only exists in the more advanced industrial nations, in various tropical countries the oil receives no treatment before being sold. The press cake so produced is often exported for compounding into animal feed, although in some cases additional yields of oil are obtained by extraction with hexane-type solvents. The peanut cakes and meal emerge as the main aflatoxin containing products, particularly as they are likely to be produced from nuts rejected for edible purposes and having a high aflatoxin level.

Aflatoxin production is not confined to growth of A. flavus on peanuts, for it has been detected in samples of cottonseed meal [83], brazil nuts and palm kernels [84] and cocoa beans [85].

The aflatoxins in low doses can produce liver cancer in experimental animals. There are differences in species susceptibility to aflatoxin carcinogenesis [86]. Most projections of the possible hazard from ingestion of aflatoxin by man are derived from experimental results [87] based on studies with male Fischer rats. These were fed on a synthetic diet in which crystalline aflatoxin B_1 was incorporated, at levels of 1-100 p.p.b. The extrapolation of the data from these experiments to man may, or may not, be valid.

Aflatoxicosis in man has been reported from India [88] where 200 villages were affected and 106 patients died. The source of infected material was maize, grown in areas affected by chronic drought. The situation in India has been reviewed by Hesseltine [89].

Production of Bacterial Toxins

Bacteria can cause problems in the large scale processing of food for if contamination occurs with a limited number of organisms, food poisoning may result. The least important numerically of these is Clostridium botulinum, which is a spore forming, strict anaerobe. This organism is the most dramatic because it produces potent protein toxins which affect the nerves of the diaphragm and pharynx causing paralysis. The toxins are not degraded by the proteolytic enzymes of the digestive tract. Clostridium botulinum is widely distributed in the soil and in the intestinal tract of herbivorous animals. The spores are resistant to heat and, if food is not properly processed, they may survive and germinate later. When the organism is grown under anaerobic conditions it produces toxins and although the incidence of the disease is low, the mortality is high. Home canned vegetables are the most frequent source of the toxin and pickled and canned meat have been implicated. Commercial canning processes are designed to destroy this particular organism.

A common type of food poisoning is caused by an enterotoxin produced by toxigenic strains of Staphylococcus aureus. This micro-organism is ubiquitous and toxigenic strains are present in ca. 25% of the population. Such carriers can infect food if they are employed as food handlers, and processed meats, sandwich spreads, potato salad and milk products provide excellent substrates. The toxaemia produced is rarely fatal, and recovery is usually complete in 24-48 hr after onset of symptoms.

The most common causes of food poisoning reported in England and Wales in 1966 were members of the Salmonellae, which were responsible for 65% of the cases. The symptoms are rather generalized with abdominal pain, chills, fever, diarrhoea and vomiting. The organisms may be spread on inadequately cooked or processed poultry or egg products.

Food poisoning cases may result from eating food contaminated with enterococci, or in meat products if Cl. perfringens spores are present. A number of outbreaks of food poisoning have been attributed to Bacillus cereus, which is ubiquitous in dust, and under warm conditions (15-50°C), the spores germinate and an enterotoxin is produced. These outbreaks were associated with cooked rice usually from Chinese restaurants and 'take-away' shops [90].

REFERENCES

1   J.M. Wood, Advan. Environ. Sci. Technol., 2(1971)39.

2   J.M. Wood, Chem. Eng. News, 1971, p. 22.

3   A. Jernelov, in R. Hartung and B.D. Dinman (Editors), Environmental Mercury Contamination, Ann Arbor Sci. Publ., Ann Arbor, Michigan, 1972, p. 167.

4   S. Kitamura, Jap. J. Hyg., 24(1969)76.

5   L. Landner, Nature, 230(1971)452.

6   J.W. Vonk and A. Kaars Sijpesteijn, Antonie van Leeuwenhoek, 39(1973) 505.

7   M. Yamada and K. Tonomura, J. Ferment. Technol., 50(1972)159.

8   K. Rissanen 1974 cited by A. Jernelov and A.L. Martin, Ann. Rev. Microbiol., 29(1975)61.

9 A.O. Summers and S. Silver, J. Bact., 112(1972)1228.

10 K. Furukawa, T. Suzuki and K. Tonomura, Agr. Biol. Chem., 33(1969)128.

11 M.K. Hamdy and O.R. Noyes, Appl. Microbiol., 30(1975)424.

12 E.S. Pattison, Science, 170(1970)870.

13 F. Challenger, Advan. Enzymol., 12(1951)429.

14 D.P. Cox and M. Alexander, Bull. Environ. Contam. Toxicol., 9(1973)84.

15 R.S. Braman and C.C. Foreback, Science, 182(1973)1247.

16 B.C. McBride and R.S. Wolfe, Biochemistry, 10(1971)4312.

17 E.M. Bautista and M. Alexander, Soil Sci. Soc. Amer. Proc., 36(1972) 918.

18 M.L. Bird and F. Challenger, J. Chem. Soc., 1939, p. 163.

19 U. Schmidt and F. Huber, Nature, 259(1976)157.

20 J.D. Nelson, H.L. McClam and R.R. Colwell, Proc. Ann. Conf. Marine Technol. Soc., 8th, Washington D.C., 1972, p. 302.

21 C. Huey, F. Brinckman, S. Grim and W.P. Iverson, Proc. Int. Conf. Transport of Persistent Chemicals in Aquatic Ecosystems, Ottawa, May 1st-3rd, 1974, 1975.

22 R.W. Fleming and M. Alexander, Appl. Microbiol., 24(1972)424.

23 E. Robinson and R.C. Robbins, in S.F. Singer (Editor), Global Effects of Environmental Pollution, Reidel Publ., Dordneckt, Netherlands, 1968, p. 106.

24 A. Abeliovich and Y. Azov, Appl. Environ. Microbiol., 31(1976)801.

25 I.C. McRae and R. Ancajas, Plant Soil, 33(1970)97.

26 R.E. Luebs, K.R. Davis and A.E. Laag, J. Environ. Qual., 2(1973)137.

27 G.L. Hutchinson and F.G. Viets, Science, 166(1969)514.

28 Committee on Nitrate Accumulation, Accumulation of Nitrate, Nat. Acad. Sci., Washington D.C., 1972.

29 N. Gruener and H.I. Shuval, in H.I. Shuval (Editor), Developments in Water Quality Research, Ann Arbor-Humphrey Sci. Publ., Ann Arbor, Michigan, 1970, p. 89.

30 R.H. Harmeson, F.W. Sollo and T.E. Larson, J. Amer. Water Works Ass., 63(1971)303.

31 L.R. Webber, in B. Westley (Editor), Identification and Measurement of Environmental Pollutants, National Research Council of Canada, Ottawa, 1971, p. 110.

32 L.D. King and H.D. Morris, J. Environ. Qual., 1(1972)442.

33 F. Selenka, Z. Bakt. Parasitkde Abt. I, 155(1971)58.

34 D.S. Curtis, Soil Sci., 68(1949)441.

35 J.L. Shank, J.H. Silliker and R.H. Harper, Appl. Microbiol., 10(1962) 185.

36 M. Tanaka, Nature, 171(1953)1160.

37 W. Verstraete and M. Alexander, J. Bact., 110(1972)955.

38 W. Verstraete and M. Alexander, J. Bact., 110(1972)962.

39 W. Verstraete and M. Alexander, Environ. Sci. Technol., 7(1973)39.

40 L.T. Delaney, H.W. Schmidt and C.F. Stroebel, Proc. Staff Meet. Mayo Clinic, 31(1956)189.

41 J.M.P. Lieb, W.N. Davis, T. Brown and M. McQuiggan, Amer. J. Med., 24(1956)471.

42 L.C. Wang and R.H. Burris, J. Agr. Food Chem., 8(1960)239.

43 B.S. Alam, I.B. Saporoschetz and S.S. Epstein, Nature, 232(1971)199.

44 R.C. Braunberg and R.E. Dailey, Proc. Soc. Exp. Biol. Med., 142(1973) 993.

45 G.M. Hawksworth and M.J. Hill, Brit. J. Cancer, 25(1971)520.

46 A. Ayanaba and M. Alexander, Appl. Microbiol., 25(1973)862.

47 A.L. Mills and M. Alexander, Appl. Environ. Microbiol., 31(1976)892.

48 M.J. Hill, J. Med. Microbiol., 5(1972)14.

49 G.M. Hawksworth, M.J. Hill, G. Gordillo and C. Cuello, in P. Bogovski and E.A. Walker (Editors), N-nitroso Compounds in the Environment. Proceedings of a Working Conference held at the International Agency for Research on Cancer, Lyon, France, 17th-20th Oct. 1973, IRAC Scientific Publications No. 9, 1975, p. 229.

50 A. Geleperin, V.J. Moses and G. Fox, Ill. Med. J., 149(1976)251.

51 E. Eriksson, J. Geophys. Res., 68(1963)4001.

52 V.J. Marchesani, T. Towers and H.C. Wohlers, J. Air Pollut. Contr. Ass., 20(1970)19.

53 D.C. Grey and M.L. Jensen, Science, 177(1972)1099.

54 L.L. Smith and D.M. Oseid, Water Res., 6(1972)711.

55 R. Rodriguez-Kabana, J.W. Jordon and J.P. Hollis, Science, 148(1965) 524.

56 M. Alexander, Adv. Appl. Microbiol., 18(1975)1.

57 J.R. Freney, Aust. J. Biol. Sci., 13(1960)387.

58 F.J. Leinweber and K.J. Monty, J. Biol. Chem., 240(1965)782.

59 K. Soda, A. Novogrodsky and A. Meister, Biochemistry, 3(1964)1450.

60 L.F. Elliott and T.A. Travis, Soil Sci. Soc. Amer. Proc., 37(1973)700.

61 W. Moje, D.E. Munnecke and L.T. Richardson, Nature, 202(1964)831.

62 J. Ruiz-Herrera and R.L. Starkey, J. Bact., 104(1970)1286.

63   R.E. Kallio and A.D. Larson, in W.D. McElroy and H.B. Glass (Editors), Amino Acid Metabolism, John Hopkins Press, Baltimore, Maryland, 1955, p. 616.

64   H. Kadota and Y. Ishida, Ann. Rev. Microbiol., 26(1972)127.

65   A.J. Francis, J. Adamson, J.M. Duxbury and M. Alexander, in T. Rosswall (Editor), Modern Methods in the Study of Microbial Ecology, Swedish National Research Council, Stockholm, 1973, p. 485.

66   R.K. White, E.P. Taiganides and G.D. Cole, in Livestock Waste Management and Pollution Abatement, Amer. Soc. Agr. Eng., St.Joseph, Missouri, 1971, p. 110.

67   J.M. Lynch, Nature, 240(1972)45.

68   B. Weinstock and H. Niki, Science, 176(1971)290.

69   R.C. Robbins, K.M. Borg and E. Robinson, J. Air Pollut. Contr. Ass., 18(1968)106.

70   W.O. Robinson, Soil Sci., 30(1930)197.

71   G.V. Dobrovol'skiy, I.P. Bab'yeva and A.P. Lobutev, Sov. Soil Sci., 2(1960)1181.

72   R.A. Lamontague, J.W. Swinnerton and V.J. Linneubom, J. Geophys. Res., 76(1971)5117.

73   D.F. Wilson, J.W. Swinnerton and R.A. Lamontague, Science, 168(1970) 1577.

74   J.R. Hennequin and C. Juste, Ann. Agron., 18(1967)545.

75   A.J. Feuell, in L.A. Goldblatt (Editor), Aflatoxin, Academic Press, New York, 1969, p. 187.

76   P.J. Brook and E.P. White, Ann. Rev. Phytopathol., 4(1966)171.

77   G. Barger, Ergot and Ergotism, Gurney & Jackson, London, 1931.

78   A.Z. Joffe, Mycopathol. Mycol. Appl., 16(1962)201.

79   D.E. Richards, in S. Kadis, A. Ciegler and S.J. Ajl (Editors), Microbial Toxins, Vol. 8, Academic Press, New York, 1972, p. 3.

80   F.A. Norstadt and T.M. McCalla, Soil Sci. Soc. Amer. Proc., 32(1968) 241.

81   P.K.C. Austwick, Br. Med. Bull., 31(1975)222.

82   J.W. Dickens and T.B. Whitaker, Peanut Sci., 2(1975)45.

83   M.E. Whitten, Cotton Gin Oil Mill Press, 67(1966)7.

84   Report of the Tropical Products Institute, 1965, H.M.S.O., London, 1966, p. 6.

85   A.J. Feuell, Soc. Chem. Ind. Monograph, 23(1966)129.

86 G.N. Wogan, in H. Busch (Editor), Methods in Cancer Research, Vol. 7, Academic Press, New York, 1973, p. 309.

87 G.N. Wogan, S. Paglialuna and P.M. Newberne, Food Cosmet. Toxicol., 12(1974)681.

88 K.A.V.R. Krishnamachari, R.V. Bhat, V. Nagarajan and T.B.G. Tilak, Lancet, (i)(1975)1061.

89 C.W. Hesseltine, Mycopathologia, 58(1976)157.

90 R.J. Gilbert, M.F. Stringer and T.C. Peace, J. Hyg., 73(1974)433.

AIR POLLUTION AND MICRO-ORGANISMS

Micro-organisms can contribute to air pollution directly by the production of gaseous pollutants (see p. 13-16), or they may themselves be
potential invaders of the upper respiratory tract where they can cause
infectious diseases. Alternatively they may be a more direct hazard and
cause allergic responses. The presence of aerosols of micro-organisms
from wastewater treatment process can also be a potential danger.

The presence of man-made pollutants in the environment may affect
micro-organisms, and essential micro-organisms required for the cycling of
nitrogen, phosphorus and other elements may be killed. It is also possible
that the interactions and balance between man and organism can be altered
adversely.

Effect of Air Pollutants on Micro-organisms

Micro-organisms can be affected by atmospheric pollution, although
little research has been carried out on these problems. Where work has
been carried out using plants there is often a reciprocal relationship
between the concentration of pollutant (C) and the length of exposure time
(T). The exposure of plants for long time intervals at low pollutant concentration can give equivalent biological responses compared to a short
exposure at high concentration i.e. C x T = constant. This relationship
may hold with micro-organisms because Watson [1] showed that if spores of
Sclerotinia fructiola were exposed to ozone, when the exposure time was
doubled, half the spores were killed at ca. 50% of the ozone concentration
normally required. Similar relationships were also observed for growing
sclerotia of Botrytis sp., Rhizoctonia tuliparum and Sclerotium delphinii
when exposed to sulphur dioxide, ammonia or hydrogen sulphide at concentrations of 1 to 1,000 p.p.m. for 1 to 960 min [2].

This toxic effect of exposure to gases has been used by many industries.
Ozone has been used to purify water [3].. It has also been used in the food
industry, in meat tenderising and storage rooms for control of the growth
of contaminating bacteria and fungi [4]. Also, ozone has been used at concentrations up to 10 p.p.m. for controlling mould growth on Cheddar cheese
[5], and to control fungal growth on bread, leather, apples and lemons.

Sulphur dioxide, and in particular metabisulphite is used in wine
making, and in the food industry as a general preservative [6], particularly
for fruit juices, and is also a recommended sterilizing agent in amateur

winemaking. The lactic acid and acetic acid bacteria are less resistant
to sulphur dioxide than yeasts; aerobic yeasts being more sensitive than
anaerobic yeasts [6].

Formaldehyde vapour has been employed for fumigation and sterilization
of closed spaces, surgical or medical equipment, blankets and wool. Its
activity, like many other agents used for gaseous sterilization (including
ethylene oxide), is influenced by relative humidity.

The effect of air pollution on lichens is important because there is
good correlation between laboratory studies and field data. The partners
in a lichen are an algal clone (the phycobiont) and a fungus or a fungal
clone (the mycobiont). The association is not obligatory for either partner. The same organisms, as species, are found in a free-living state and
both the algal and fungal partners in a lichen can be cultivated in artificial media as pure cultures. However, lichens will grow in habitats
which do not support either partner alone, which suggests that both partners obtain some benefit from the association. The bulk of a lichen is
fungal protoplasm. Each association has a particular morphology, pigmentation and habitat. The lichen may be crustose (developing as a thin crust
on the surface of rocks), foliose (flattened, often ribbed) or fruticose
(attached at one point and growing away from the point of attachment, like
a shrub in appearance).

Lichens can be used as very sensitive indicators of sulphur dioxide
pollution. Most of the data available is on sulphur dioxide pollution and
there is little information on other pollutants. If the situation around
a shale oil works or an iron sintering plant is considered then three
zones can be observed. These are characterized by an absence of lichens,
an area of scant lichen growth and an area with normal growth. These zones
can be correlated with the sulphur dioxide level in the air, the pH and
sulphate level of the soil and surface water [7]. The observed effects
are due to direct poisoning of the lichens by the sulphur emissions and to
an alteration of the substrate.

The absence of lichens has been reported in many parts of the world
including Port Talbot [8], Newcastle-on-Tyne [9] and Brooklyn, N.Y. [10].
Pyatt [8] showed that there was also a change in the species which were
found as one moved from the city centre i.e. there was a transition from
crustose to foliose to fructicose. Fenton [11] considered that crustaceous
lichens were resistant to atmospheric pollution because they had a slow
growth rate and had only one surface exposed to the atmosphere.

The primary target for the action of sulphur dioxide appears to be the algal symbiont, but the fungal symbiont is also affected adversely. Exposure of several lichens to 5 p.p.m. for 24 h of sulphur dioxide caused permanent plasmolysis of the algal cells with brown spots on the chloroplasts and the chlorophyll was bleached [7]. The chlorophyll a content of the algal symbiont was probably destroyed.

LeBlanc et al. [12] showed that the volatile fluorides Na_3AlF_6 and SiF_4 and hydrogen fluoride which were released from an aluminium factory also adversely affected the lichen flora. Morphological changes were produced with plasmolysis of the algal cells and changes in pigmentation. Fluoride is accumulated by lichens and initially chlorophylls are degraded, and then β-carotene and the xanthophylls are destroyed [13]. The fluoride ion is also a potent inhibitor of the enzyme enolase and as such could effectively block glucose metabolism.

If there is a decrease in the numbers of lichens there will be a decrease in the numbers of herbivores which feed on them. Similarly if heavy metals are accumulated by lichens, then these metals will accumulate in the herbivores, and so the diversity of species will probably be restricted.

The morphology of micro-organisms can be changed after exposure to atmospheric pollutants. Lantzch [14] showed that when Bacillus oligocarbophilus was grown under autotrophic conditions in a carbon monoxide atmosphere it was seen as a thread-like network. However, if it was grown on ordinary laboratory media then it produced a coccal (spherical) form.

Ozone sensitive fungi will restrict the production of aerial hyphae and Trichoderma viride grows close to the agar [15]. Pigmented microbial cells may be more resistant to ozone than non-pigmented cells, because a carotenoid strain of Mycobacterium carothenum was more resistant than a white mutant [16].

Air pollutants can also affect microbial interactions with higher organisms. In regions where the level of sulphur dioxide was high, various fungi, which are pathogenic to plants, were either eliminated or reduced. For example, blister rust of eastern white pine [17], oak mildew [18] and wheat stem rust [19] are reduced in such areas. However, there have been reports of the proliferation of certain fungi in regions where the level of sulphur dioxide is high [19,20]. The black spot disease of roses is eliminated in areas with a high level of atmospheric ammonia [21].

Other forms of air pollution can affect fungal growth, for Schoenbeck [22] reported that the dust from a cement mill was responsible for a higher incidence of leaf spot disease on sugar beet. Pollutants may have a beneficial effect in certain instances e.g. sulphur dioxide emissions have satisfied the sulphur requirements of plants, but not micro-organisms, in sulphur deficient soils [23].

The cycling of various elements is dependent on the activity of micro-organisms, and air pollutants can affect these cycles. Nitrogen fixation which occurs in the nodules of leguminous and other plants is inhibited by as little as 0.01% carbon monoxide in legumes [24], and a similar response is found in red clover [25]. Ozone can also affect the production of Rhizobium nodules in Pinto bean plants [26] and soya bean plants [27].

The nitrogen cycle was affected by airborne arsenic, present as arsenic trioxide, which was produced by copper smelters and deposited in the soil. Wullstein and Snyder [28] showed that the soil from around a smelter had an arsenic concentration of 110 to 250 p.p.m. and the rates of conversion of amino acids to ammonium, nitrite and nitrate were decreased when compared to uncontaminated soil. The organisms which were mainly affected were the nitrifiers.

The distinction between a pathogenic organism and a non-pathogenic organism is somewhat arbitary. Any organism capable of a parasitic mode of life is a potential pathogen. The only distinguishing feature between a pathogen and a non-pathogen is the response of the host; pathogenicity is not really a property of the organism but of the host/organism interaction. This interaction, apparently, can be affected by air pollutants because there is evidence that the incidence of respiratory infection can be altered by interference with the body's defence mechanism, so increasing susceptibility to disease. Miller and Ehrlich [29] showed that hamsters and mice which were exposed to 0.8 p.p.m. ozone for 100 h, or 4.4 p.p.m. for 3 h had a decreased resistance to infection produced by aerosols of Streptococcus sp. Also, if mice were exposed to irradiated car exhaust fumes for 11 h then exposed to a streptococcal aerosol, they showed enhanced mortality [30]. Emik et al. [31] demonstrated that pneumonitis was more prevalent in mice exposed to California air than when they were exposed to filtered air. When the level of nitrogen dioxide in the air was 2 to 3 p.p.m. the susceptibility and mortality of mice to an aerosol challenge of Klebsiella pneumoniae was increased [32].

There may be an increase in the resistance of the host to microbial infection because of the stimulation of macrophage phagocytic activity by certain pollutants [33]. When mice were exposed to 100 p.p.m. of carbon monoxide, for up to six days, there was an increase in resistance to

infection by <u>Listeria monocytogenes</u> when compared to mice exposed to air.
If mice were pre-infected with influenza virus and then exposed to sulphur
dioxide for seven days, the concentration ranging from 2.9 to 34 p.p.m.,
there were fewer cases of pneumonia at concentrations up to 7 p.p.m., and
above this value there was a greater incidence of pneumonia [34]. This
suggests that the effects of gaseous pollutants may not be clear cut and
graded responses are more likely.

The presence of dust in the atmosphere may have some effect on infections. Exposure to aerosols of silica dust, feldspar dust [35], aluminium
dust [36] or bituminous coal dust [37] gave rise to less lobar pneumonia
in rats, after infection with type I <u>Pneumococcus</u>, than in the control
group which had not been exposed to the dusts. On the other hand, cement
dust had no effect on the incidence of similar infections in rats [38].
The effect of dusts seems to be primarily on pulmonary bactericidal activity and not on pulmonary mucous elimination.

There may be an interaction between atmospheric pollutants and viruses
which could give rise to the production of cancerous cells. Kotin and
Wiseley [39] used a strain of black mice C57, which had a low incidence of
spontaneous lung carcinoma development and a high resistance to lung tumour
induction. These mice were exposed to petrol fumes which had been reacted
with 1 to 2 p.p.m. of ozone. Some of the mice were also simultaneously
infected intranasally with PR8 influenza virus, V strain of influenza virus
B, and Sendai newborn pneumonitis virus. Of the groups studied, only the
mice which were exposed to the fumes and virus developed squamous carcinoma
of the lung. The control groups which were exposed to fumes, or to virus
alone, did not develop carcinoma. Virus infected cells grown in tissue
culture, can be transformed by exposure to city air which contains benzo-
(a)pyrene and can then cause tumours when transplanted into mice [40].

The effects of air pollutants can be more damaging than the immediate
expected illness produced by the chemical alone. The effect of such pollutants on micro-organisms at the cellular and subcellular level, bringing
about changes in enzymes, or by producing mutations or selecting resistant
mutants could give rise to long term unexpected effects.

Micro-organisms as Air Pollutants

Micro-organisms which are present in the air can be pollutants. Pollutant organisms include bacteria, viruses, spores of lichens and fungi,
small algal and protozoal cysts. These organisms may be involved in the
production of infectious disease or in eliciting allergic responses.

Many human disease can be transmitted by the aerial route, particularly by droplet infection in crowded conditions. There are many bacterial and fungal infections which can be acquired by this route and these are summarized in Tables 1 and 2. Viral infections, including influenza, the common cold, poliomyelitis, measles and smallpox can be spread by droplet infection. There are also similar disease conditions which can occur in cattle, including bovine rhinotracheitis, pulmonary pasteurellosis, foot and mouth disease, rinderpest and bovine contagious pleuro-pneumonia. These disease conditions can lead to economic loss, as do similar diseases in poultry e.g. Merek's disease, avian infectious bronchitis and Newcastle disease [41].

The presence of certain organisms in a working environment can constitute an occupational hazard. Fungus spores are almost always present in air, but their numbers and types will vary depending on the time of day, season, location and humidity [42,43]. This is particularly true on farms and is correlated with man's activity. Most of the spores will cause no trouble, but some people become sensitized and develop allergic reactions.

Allergy has been defined as the acquired, specific, altered capacity to react to a stimulus [44]. It can be acquired only by exposure to an allergen, is specific to that allergen, and is characterized by a reaction which is not present before sensitization [45]. The allergic reactions to mouldy material usually occur in the respiratory system and are due, mainly, to fungus and actinomycete spores. The spores and fragments of hyphae from species of Penicillium, Mucor, Rhizopus, Candida, Aspergillus, Fusarium, Helminthosporium, Alternaria and Cladosporium are capable of being allergenic [46].

The chemical composition of the inhaled particles will affect its antigenicity and the size of the spore influences the distance it will travel in the respiratory system before it is deposited. During respiration particles which are larger than 10 μm are deposited in the nose and pharynx, with those smaller than 5 μm (with an optimum of 2 to 4 μm) reaching the alveoli [47] (Table 3).

The allergic response produced by the allergen may be immediate (Type I reaction) [48], or may be delayed for several hours after exposure to the allergen (Type III). The spores which are associated with Type I allergy are usually larger than 5 μm, whilst those concerned with Type III allergy are usually smaller. The Type III reaction is associated with the disease condition described as extrinsic allergic alveolitis or hypersensitivity pneumonitis (Table 2).

(Continued on p. 38)

Table 1.

BACTERIAL INFECTIONS WHICH MAY BE ACQUIRED BY INHALATION

Disease	Organism
Pulmonary tuberculosis	Mycobacterium tuberculosis
Pulmonary anthrax	Bacillus anthracis
Staphylococcal respiratory infections	Staphylococcus sp.
Streptococcal respiratory infections	Streptococcus pyogenes
Pneumococcal pneumonia	Diplococcus pneumoniae
Nocardiosis	Actinomadura asteroides
Q fever	Coxiella burnetii
Whooping cough	Bordetella pertussis
Diphtheria	Corynebacterium diphtheriae
Sinusitis, bronchitis	Haemophilus influenzae
Primary atypical pneumonia	Mycoplasma pneumoniae
Pneumonic plague	Yersinia pestis

Table 2.

FUNGI AND ACTINOMYCETES ASSOCIATED WITH RESPIRATORY INFECTION

Disease	Source	Organism
Extrinsic Allergic Alveolitis		
Farmer's lung	Mouldy hay	Micropolyspora faeni Thermoactinomyces vulgaris
Bagassosis	Mouldy sugar cane	T. vulgaris
New Guinea lung	Mouldy roofing thatch	Streptomyces olivaceous
Maple bark pneumonitis	Mouldy maple bark	Cryptostroma corticale
Malt worker's lung	Mouldy malt	Aspergillus fumigatus
Mushroom worker's lung	Mushroom compost	M. faeni T. vulgaris
Sequoiosis	Mouldy redwood sawdust	Graphium and Pullalaria
Other Infections		
Cryptococcosis	Pigeon droppings	Cryptococcus neoformans
N. American blastomycosis	Soil, probably restricted distribution	Blastomyces dermatitidis
S. American blastomycosis	Saprophyte on vegetation or soil	Paracoccidiodes brasiliensis
Coccidioidomycosis	Soil	Coccidioides immitis
Histoplasmosis	Chicken or bat droppings	Histoplasma capsulatum
Sporotrichosis	Straw, sphagnum moss	Sporothrix schenckii
Adiaspiromycosis	Nests of field mice	Emmonsia crescens

Table 3.

THE RELATIONSHIP BETWEEN THE PARTICLE SIZE OF INHALED SPORES
AND PENETRATION OF THE RESPIRATORY TRACT

Part of Respiratory Tract	Size of Particles which Penetrate (μm)
Nasal cavity	Over 60
Trachea	30-60
Primary bronchus	20
Secondary bronchus	10
Terminal bronchus	6
Respiratory bronchiole	4
Alveoli	Below 3

In Britain, farmer's lung is the commonest variety of extrinsic allergic alveolitis and is caused by spores of Micropolyspora faeni and Thermoactinomyces vulgaris. There are differences in regional prevalence which may depend on variation in rainfall and agricultural practice. The highest incidence is in the wetter parts of Scotland [49]. The storage of hay with a high water content encourages the growth of mould, and of the thermophilic actinomycetes which are distributed in the atmosphere and inhaled when the hay is disturbed and fed to cattle. The spore concentration may reach 1.7×10^9 per m^3 when mouldy hay is shaken [50] or 2.9×10^9 per m^3 when grain silos are unloaded [51]. The peak incidence of farmer's lung is during the winter months following a wet summer. Other cases may follow when mouldy hay is cleared out from a byre or barn. The risks are greater for the small scale, poorer farmer, who may have to collect hay in conditions which are not ideal and who stores it in badly ventilated buildings.

Extrinsic external alveolitis was reported [52] in four office workers, where the causal agent was a thermophilic actinomycete which had contaminated an air conditioning plant.

The allergenic spores found in fodders which cause Type I allergy are Cladosporium herbarum, Alternaria terius and Aspergillus fumigatus. The first two are common field fungi, whilst A. fumigatus increases on storage and can be found in the air over cereals stored in a silo [53]. Fungal airborne allergens which have been associated with asthma, include Epicoccum, Cladosporium and Alternaria. Some individuals are sensitive to the inhalation of Leptosphaeria-type ascospores and spores of the mirror yeasts and toadstools [54].

Many micro-organisms, including viruses, bacteria, algae, protozoa and yeasts, lack dispersal mechanisms. These organisms rely on external mechanical forces for their liberation into the atmosphere. Animal viruses are disseminated during talking, coughing or sneezing by infected persons, or after shaking bedding used by infected persons [55]. Plant viruses and bacteriophage may enter the airborne state on water droplets or rafts of debris [56].

Wind, animal activity, and the mechanical disturbance of dust by sweeping, can send dust particles containing bacteria into the atmosphere [56]. The splashing of rain can give rise to aerosols, containing bacteria or fungi, which can spread plant or human pathogens.

A possible health hazard which may result from aerosol production comes from the use of vapour- or aerosol-type room humidifiers. Bacteria are emitted from both types with <u>Pseudomonas aeruginosa</u> predominating, but other bacteria can be present e.g. <u>Ps. algicaligenes</u>, <u>Enterobacter</u> and <u>Herella</u> [57].

The filamentous fungi are best adapted for the aerial transmission of spores. The discharge can be passive, which means adaptation for distribution by external forces, e.g. wind or raindrops. In this category are the <u>Aspergilli</u>, <u>Penicillia</u> and <u>Cladosporia</u> which produce dry spores distributed by air turbulence. In other fungi, the spores are found within a slime and require the splashing of raindrops for distribution. The mechanism involved may be more complex e.g. in puff-balls (<u>Lycoperdon</u>) or <u>Cyathus striatus</u>, but raindrops are still required for dispersal. In the second category are those which use a violent discharge mechanism, the energy being supplied by the fungus. These ascomycetes require changes in the turgour pressure within the ascus to cause the ascus to burst, and release the ascospores. Many basidiospores are released in basidiomycetes by bursting of a gas bubble.

Aerosols from Wastewater Treatment Processes

The problem involves deciding to what extent viable aerosols, produced by wastewater treatment processes, represent a real health risk. Particles which are in the medically important range, 1 to 5 μm, can be generated by many modern wastewater treatment processes which use stirred or aerated tanks and so produce aerosols. Human pathogens are present in wastewater, possibly in large numbers, at all stages of handling and if aerosols are produced, they may be carried large distances by wind currents. With the increase of urban development in recent years, large populations are now located near treatment plants, so increasing the risk. This makes a full understanding of the problems involved of greater importance.

The simplest method of sampling air for viable micro-organisms is to use agar settling plates. These consist of nutrient agar, or selective media, in petri dishes which are exposed to the atmosphere and the viable particles which fall on them can be counted after incubation. The viable particles can be measured i.e. those which contain at least one viable cell, in each unit of time and in a unit of agar surface. Glycerin-coated swabs collect viable particles by the same mechanism [58]. Agar plates can be angled to the wind direction and can measure the viable particles which are impacted on to the agar surface [59]. The results obtained by these methods are not directly related to the concentration of viable organisms per unit volume of air.

The solid media impactors e.g. the Wells air centrifuge [60], the Andersen drum [61] and six-stage impactor samplers [62] measure the viable particles per unit volume of air sampled, subject to the limits of their efficiencies. The Andersen six-stage impactor is the most popular sampling device. If a mono-layer of oxyethylene docosanol is used, then the problems associated with drying of the agar surface can be overcome [63]. This sampler can also distinguish the particle size spectrum of an aerosol containing viable organisms, whilst the drum sampler will determine the variation in the concentration of viable particles with time. The viable particles recovered by the samplers which use solid media will give a single colony after incubation. However, one viable particle may contain one or several viable organisms.

When liquid impingers are used, the viable particles in each unit volume of air can be estimated, but the particles may break up on collection in the liquid medium. The number of colonies produced after incubation are usually greater in number than the number of viable particles originally present.

The count obtained by any method should be recorded as the number of viable particles, not as the number of organisms, because each particle may contain more than one organism. It should also be noted that the medium used and the temperature selected will not allow the growth of all micro-organisms present in any one sample. The results obtained will therefore be an under-estimation of the true number of viable particles per unit volume.

When a sampling method is chosen it must have a reasonable efficiency in the size range which is of medical significance i.e. 1-10 μm diameter. It must be able to measure the concentration of viable particles per unit volume of air, permit the identification of the organisms, minimise the logistic problems encountered in sampling and, wherever possible, give a measurement of the size range of the particles.

There have been a few comparative studies made of the various samplers currently used. The viable particle recovery levels of the Wells air centrifuge and the Andersen six-stage impactor were comparable, in laboratory studies [64]. The midget impinger [65] gave results which were 31 times higher than these samplers. In field sampling tests, where all three samplers were used together, the results were very similar. Higgins [66] regarded these results as being characteristic of an aerosol which contained a large number of viable organisms in each particle. This could have been because, in this particular study, the impingers were placed 6.4 m from the edge of a trickling filter during sampling and could have been collecting unevaporated spray.

In another study, the Andersen six-stage impactor, the all-glass Millipore impinger and agar settling plates were used [67]. This involved sampling for airborne bacteria downwind from two activated sludge aeration tanks and one grit chamber at two separate treatment plants. Samples were taken 0.6 m above the liquid surface. The Andersen six-stage impactor recovered 870 colonies/cu. ft of air, the impinger 1,170 colonies/cu. ft and agar plates 220 colonies/cu. ft. The settling plates recovered as many different groups of Enterobacteriaciae as did the Andersen impactor, but not as many different types of other enteric organisms.

These studies illustrate the difficulties encountered in sampling air for the presence of viable organisms and explain why it is difficult to correlate results obtained by two different methods. Most workers use only one method and assume that it is 100% efficient.

The generation of viable bacterial aerosols from simulated wastewater which had been inoculated with coliforms, Streptococcus sp., Serratia marcescens and Bacillus subtilis spores was examined under laboratory conditions [66]. The aerosols were produced by pumping air through sintered glass diffusers which were submerged in the inoculated water. Air samples were taken downwind in a wind tunnel and viable particles collected on agar settling plates by the Andersen six-stage impactor and by a slit sampler [68]. Isokinetic sampling conditions were maintained with the slit sampler and the Andersen six-stage impactor by using glass sampling probes up to 31 mm. in diameter inserted into the wind tunnel. This possibly discriminated against the larger diameter particles. There were no effects which could be attributed to air or liquid temperature, or relative humidity variations. The size distribution of the aerosol particles and the size distribution were not dependent on bubble diameter. Under comparable conditions the recovery of S. marcescens was up to 5 times that of B. subtilis. If there was loss of viability of S. marcescens during aerosol production, it could have been present in aerosols at 20 times the level of B. subtilis. The recovery of coliforms was very low, with Escherichia coli and E. freundii not being recovered and E. aurescens being recovered at less than 1% of B. subtilis. Klebsiella aerogenes was recovered only from large unevaporated droplets. The production of viable aerosols may be species selective. Serratia marcescens probably concentrates at the surface and coliforms in the bulk of the liquid. From these results it appears that coliforms may not be a suitable indicator organism, and the total viable count may be more significant.

A similar result with S. marcescens was obtained by Blanchard and Syzdek [69] who found that the level in bursting bubbles was 10 to 1,000 times higher than in the liquid from which the drops were formed. The level of coliform organisms was higher in the thin films of surface water compared to sub-surface water and this population could be transferred to the atmosphere when bubbles burst [70].

The production of aerosols from bubbles bursting at the surface of aerated liquid inoculated with 10^7-10^8 spores/ml of B. subtilis var. niger was dependent on the aeration bubble size, the concentration of spores in suspension and the composition of the liquid [71]. These laboratory studies suggest that the aeration of contaminated wastes and other liquids could be responsible for the production of infective aerosols.

The earliest study of the airborne spread of bacteria from wastewater was carried out by Horrocks [72]. He showed that when wastewater flowing in sewers was seeded with Bacillus prodigiosus, this organism could be recovered on agar settling plates placed in branch ventilation risers 15 m above the flowing wastewater. This was so, even if the particular branch risers were upstream from the flow of seeded wastewater. The organism was not recovered on control plates, or from any branches which were separated from the seeded wastewater by running traps. Salmonella typhi and coliform organisms were also collected on agar settling plates which were placed in branch risers. The organisms may be injected into the air when bubbles burst at the surface of the water, or by re-entrainment of dried particles from the walls of the sewers, or by ejection of droplets from the flowing wastewater. The main conclusion from this work was that air inlets to a sewer system could be a potential source of danger if they were placed near ground level.

Further investigations were carried out on the total numbers of organisms and coliform bacteria present in the air near activated sludge aeration tanks and trickling filters using a Wells air centrifuge [73]. There were 22.3 organisms/cu. ft at the sources and up to 3.9 organisms/cu. ft at a distance. The respective coliform counts were 2 organisms/cu. ft and less than 1 organism/cu. ft. There were no conclusions made about the health hazards which might arise from these levels. Attention was drawn to the considerable dilution of the contaminated air which could be significant in reducing the hazard.

The airborne spread of bacteria downwind from a trickling filter plant was studied using the Andersen six-stage impactor sampler, the Wells air centrifuge and the midget impinger [64]. All the samplers were placed on the ground during sampling. The main source of bacterial aerosols was the trickling filter; whilst the inflow, sedimentation tanks, sludge digester and sludge drying beds were only minor contributors. The airborne levels of total bacteria, Escherichia coli and Klebsiella aerogenes were estimated to a distance of 90 m downwind from the filters and controls were placed upwind. Coliform organisms were recovered 30 m downwind from the filters. The maximum coliform recoveries, at a distance greater than 0.9 m from the edges of the filters, were 6 E. coli and 5 K. aerogenes/cu. ft. The total number of bacteria were up to 144 viable particles/cu. ft at the edge of one filter and 80 viable particles/cu. ft at 15.2 m downwind. There was no E. coli or K. aerogenes in the upwind controls, but in 4 out of 18 samples the upwind control had counts equal to, or higher than, the downwind recoveries. In this work, the relative humidity, air temperature, wind direction and speed were all recorded. No correlation could be shown between the spread of coliform organisms in the air and the relative humidity or air temperature. There was a positive relationship between the updraught occurring over the trickling filters, wind speed and the recovery of E. coli. The conditions of filter ventilation are probably related to the production, range and spread of airborne bacteria which are emitted from the filter, and the wind probably determines the distance which the organisms travel.

The downwind dispersion of bacterial aerosols from activated sludge aeration tanks was examined using agar settling plates [74,75]. The plates were held 1.2 m above the ground for 3 min with the agar surface facing into the wind at a 30° angle. The samples were taken at 6, 15.2 and 30.5 m downwind, with a control at 6 m upwind. The total count of viable organisms was 32 colonies at 6 m and 6 colonies at 30.5 m downwind from the tanks. The coliform count decreased from 5 to 1 colony at the same distances. These results were obtained by subtracting the counts obtained on the control upwind plates from the test downwind plates. The half-life for the total airborne bacteria was 0.38 s at 6 m downwind from the tanks with the half-life for the coliform organisms being less than this [74,75]. The short half-life is probably due to die off and dispersion of the organisms.

The Andersen six-stage impactor was used to study the production and downwind spread of coliform organisms from several units [76] in four activated sludge and high-rate trickling filter plants. The samples were collected for 5 min on eosin methylene-blue (EMB) agar (see chapter on water testing) and the total count obtained was taken as being coliform organisms. The samples were taken at the primary sedimentation tanks (of the activated sludge tanks) and at the primary and final sedimentation tanks and filters (of the trickling filter plants). There were twice as many organisms produced by the trickling filters as the other units of the trickling filter plant. The main source of bacterial aerosols in the activated sludge plant was from the activated sludge aeration tanks. There was a wide range in the numbers of organisms recovered from the main aerosol producers e.g. the numbers at 45 m downwind from the activated sludge aeration tanks were 1 and 154 colonies. When similar 5 min samples at 45 m downwind from the trickling filters were taken 15 and 149 colonies were obtained. However, there were differences between the four plants in the quantity and quality of the wastewater and in the weather conditions during sampling. About 50% of the particles produced were less than 5 μm in diameter and so could present a health hazard.

The Andersen drum sampler and the Andersen six-stage impactor were used in the examination of aerosols produced by a pre-aeration tank, a trickling filter and an activated sludge tank [77]. The protein-bearing aerosols, as distinct from viable particles, were estimated using a sequential membrane filter sampler. Control samples were collected upwind and the test samples were taken downwind from the various units. The downwind recoveries were up to 351 organisms/cu. ft of air sampled at the pre-aeration units, 146 organisms/cu. ft at the activated sludge unit and 77 organisms/cu. ft at the trickling filter. The upwind control count ranged from 28 to 86 colonies/cu. ft and these values were subtracted from the appropriate downwind samples. The organisms recovered from the downwind samples, when examined on selective media, contained 53 coliform colonies and 64 β-haemolytic colonies from the 180 colonies examined. Protein material was recovered from the downwind samples of the activated sludge and pre-aeration tanks. The rate of production of viable particles depended on the air temperature, wind speed, relative humidity and flow rate of the wastewater.

When cultures of B. subtilis var. globigii were released into sewers, as far as 8 km from one plant, viable organisms were recovered 5 h later from the air downwind from the pre-aeration tank. The maximum recovery level was 15 colonies/cu. ft of air. No viable organisms were recovered 10 h after release and none of this particular species was recovered from upwind controls.

Wastewater may be reused for agricultural purposes and can give rise to viable aerosols during its application. Merz [78] studied the problem when treated wastewater was sprinkled onto a golf course. The air was sampled downwind from the sprinkler and the numbers of coliform organisms determined. Coliform organisms were recovered only downwind from the sprinkler and only when close enough, at 41 m, to feel the spray. From this it was concluded that any hazard was associated only with direct contact with unevaporated droplets of water.

The literature on airborne organisms arising from the application of wastewater to land has been reviewed by Sorber et al. [79,80]. Viable coliform organisms were recovered from wastewater being discharged onto crops [81] and the viable aerosol could be carried 400 m downwind by a 5 m/s wind [82]. When an examination was carried out on settled wastewater discharged from sprinklers, bacteria were recovered 650 m downwind and a 1,000 m zone was recommended, for safety purposes, around such installations [83]. Contrary to expectation, low pressure sprays for land application may increase wastewater aerosol production [84].

When samples were taken near aeration tanks large numbers of indole-positive Klebsiella were recovered [67]. These are known respiratory pathogens and any person standing 1.5 m downwind from an aeration tank might inhale one viable Klebsiella every two breaths. This makes the assumption that each viable particle contains one Klebsiella. When the total bacteria recovered were examined 10.5 % were potential pathogens of the respiratory tract i.e. Klebsiella, Aerobacter or Proteus. Even when long term sampling was carried out no viable Shigella or Salmonella were recovered. The capsulated organisms like Klebsiella are more likely to survive in the air than non-capsulated organisms like E. coli. The survival rate of the Enterobacteriaceae was found to be only 13% of that of the total bacterial aerosol.

There was a much larger number of enteric bacterial genera and species recovered downwind compared to those recovered upwind from wastewater containing 1.2×10^{10} organisms per ml [84]. This means that the true source of the organisms was probably the aeration tanks which were being studied. In fact, Staphylococcus aureus was recovered five times more frequently downwind as upwind, whilst Alcaligenes faecalis was recovered downwind but never from upwind samples. This latter organism could be a good indicator for the production of airborne bacteria from wastewater.

In other studies [58] <u>Salmonella</u> and <u>Mycobacterium</u> were recovered from liquid wastewater in an enclosed aerated wastewater tank which was ventilated with forced draught blowers. Also, <u>Klebsiella</u> and haemolytic <u>Streptococci</u> were recovered from air sampled inside the building. The air inside the exhaust stack was found to contain <u>Klebsiella</u> and <u>Mycobacterium</u>. There were none of these organisms recovered from upwind air samples. The acid-fast organisms which were recovered in the air-stream of the stack at 22 m, produced lesions in the guinea pig. Even when ozone at 0.3 to 0.6 p.p.m. was added to the exhaust air there was no change in the levels of bacterial recovery.

The health hazard of working at or near a wastewater treatment plant has received some attention, although the conclusions are not clear cut. Coliform organisms were recovered from the spray of brush aerators at an activated sludge plant and it was concluded that a definite health hazard was present [85]. The Final Report of the American Public Health Association Committee on Standard Methods for the Examination of Air [86] stated that the number of bacteria in the air does not seem to be a factor of any great significance but may be of special interest in the dairy industry. Winslow [87] concluded that bacteria which could be emitted into the air by splashing wastewater were present in such low numbers that it ought to be possible to breathe sewer gas for 24 hr without being infected.

The calculations based on the recovery of <u>Klebsiella</u> organisms near an activated sludge aeration tank [67] suggest that an employee could inhale a maximum of 105,000 <u>Klebsiella</u> in a month. This was thought to give a very minimal risk of infection because it required the inhalation of 850,000 <u>Klebsiella pneumoniae</u> to produce an infection in squirrel monkeys and 730,000 failed to give any infection [87].

The only direct study of the effects on the level of infection in persons working at a wastewater treatment plant are somewhat inconclusive [75]. The incidence of pneumonia in 287 employees of a large Texas wastewater treatment plant was the same (at 0.002 per year per employee) as in a control group taken from 383 employees of drinking water treatment plants. However, the incidence of influenza was higher by 50% and colds by 28% in the wastewater workers. There were, however, some problems with diagnosis and reporting of the less severe illnesses.

The spraying of wastewater for irrigation may constitute a public health hazard because of the inhalation of pathogenic organisms, including viruses, in the aerosols. The disinfection processes used to treat wastewater could be less effective against pathogenic viruses than they are against the organisms which are commonly used as indicators. The health hazards of using wastewater for irrigation were reviewed by Sepp [88] and he cites many examples of infection of food and fodder following irrigation with wastewater. He also points out that aerosol production during spraying is recognized in the legislation of certain countries, for example in Eastern Europe the spraying of wastewater is usually prohibited when strong winds are blowing. There may also be requirements of chlorination of wastewater and safeguards may be needed near dwellings and waterworks.

There may be special hazards arising from the source of the wastewater, for tubercle bacilli can be found in the systems of several towns which have tuberculosis clinics [89]. This could present a hazard to workers following inhalation of viable particles. Similar special circumstances may be important when hospital waste is processed or during epidemics. There may also be an allergenic hazard because the aerosols may not contain viable bacteria, but they can contain protein material [90]. There is no information concerning the possibility of the spread of viral infections although viruses are found in wastewater [91]. Coliphages have been found in significant numbers in the air near activated sludge plants and trickling filter plants [92] and so other viruses are probably released as well. The level of coliphage could be used as an indicator of the probability of other viruses being present.

Wastewater treatment processes were considered as one of many producers of viable aerosols [93], but the airborne transmission of infection was regarded as being limited to indoors and to confined outdoor spaces. There is a report of tularaemia infection from breathing aerosols generated during beet washing operations [94]. The production of aerosols by many microbiological techniques, in particular centrifugation, can create a health hazard in the laboratory [95].

The inhalation of low concentrations of pathogens may give a low degree of immunity [96]. The exposure of treatment plant workers and populations resident near to wastewater treatment plants may produce resistance in this way and no adverse health effects will be produced. If this is the case, then only populations which are exposed occasionally to infective aerosols would be adversely affected. On the available evidence the health hazards arising from aerosols produced by wastewater treatment processes may be regarded, tentatively, as insignificant.

REFERENCES

1   R.D. Watson, Ozone as a Fungicide, Ph.D. Thesis, Cornell University, Ithaca, New York, 1942.

2   S.E.A. McCallum and F.R. Weedon, Contrib. Boyce Thompson Inst., 11 (1940)331.

3   W. Stumm, J. Boston Soc. Civil Eng., 45(1958)68.

4   R. Nagy, in Ozone Chemistry and Technology, Advances in Chemistry, Vol. 21, American Chemical Society, Washington D.C., 1959, p. 57.

5   C.A. Gibson, J.A. Elliott and D.C. Beckett, Can. Dairy Ice Cream J., 39(1960)24.

6   M.A. Joslyn and J.B.S. Braverman, Adv. Food Res., 5(1954)97.

7   D.N. Rao and F. Le Blanc, Bryologist, 70(1967)141.

8   F.B. Pyatt, Environ. Pollut., 1(1970)45.

9   O.L. Gilbert, New Phytol., 69(1970)605.

10 I.M. Brodo, Bryologist, 69(1966)427.

11 A.F. Fenton, Ir. Nat. J., 13(1960)153.

12 F. Le Blanc, G. Comeau and D.N. Rao, Can. J. Bot., 49(1971)1691.

13 T.H. Nash, Bull. Torrey Bot. Club, 98(1971)103.

14 K. Lantzch, Z. Bakt. II, 57(1922)309.

15 C.R. Hibben and G. Stotzky, Can. J. Microbiol., 15(1969)1187.

16 V.P. Zobnina and E.A. Morkovina, Microbiology, 41(1971)79.

17 S.N. Linzon, J. Air Pollut. Control Assoc., 21(1971)81.

18 G. Koeck, Z. Pflanzenkr. Pflanzenschutz, 45(1935)1.

19 A.S. Heagle, Ann. Rev. Phytopathol., 11(1973)365.

20 T.C. Scheffer and G.G. Hedgcock, USDA Technical Bulletin, 1117, 1955.

21 P.J.W. Saunders, Lichenologist, 4(1970)337.

22 H. Schoenbeck, Ber. Landesanstalt für Bodennutzungsschutz, 1(1960)89.

23 G. Stotzky and A.G. Norman, Arch. Mikrobiol., 40(1961)370.

24 C. Bond, J. Exp. Bot., 11(1960)91.

25 C.J. Lind and P.W. Wilson, Bot. Gaz., 108(1946)254.

26 W.J. Manning, W.A. Feder, P.M. Papia and I. Perkins, Environ. Pollut., 1(1971)305.

27 R.A. Reinert, D.T. Tingey and C.E. Koons, Agron. J., 63(1971)148.

28 L.H. Wullstein and K. Snyder, in H.M. England and W.T. Berry (Editors), Proc. Second Int. Clean Air Congr., Academic Press, New York, 1971, p. 295.

29 S. Miller and R. Ehrlich, Bact. Proc., Abstr. No. M123(1958)94.

30 D.L. Coffin and E.J. Blommer, Arch. Environ. Hlth., 15(1967)36.

31 L.O. Emik, R.L. Plata, K.I. Campbell and G.L. Clarke, Arch. Environ. Hlth., 23(1971)335.

32 M.R. Purvis and R. Ehrlich, J. Infect. Dis., 113(1963)73.

33 R.F. Ungar, Ph.D. Thesis, Oklahoma University, Norman, Oklahoma, 1972.

34 G.A. Fairchild, J. Roan and J. McCarroll, Arch. Environ. Hlth., 25 (1972)174.

35 A.M. Baetjer and F.J. Vintinner, J. Ind. Hyg. Toxicol., 26(1944)101.

36 F.J. Vintinner, A.M.A. Arch. Ind. Hyg. Occup. Med., 4(1951)217.

37 F.J. Vintinner and A.M. Baetjer, A.M.A. Arch. Ind. Hyg. Occup. Med., 4(1951)206.

38 A.M. Baetjer, J. Ind. Hyg. Toxicol., 29(1947)256.

39 P. Kotin and D.V. Wiseley, Prog. Exp. Tumor Res., 3(1963)185.

40 J.M. Rhim, H.Y. Cho, L. Rabstein, R.J. Gordon, R.J. Bryan, M.B. Garner and R.J. Heubner, Nature, 239(1972)103.

41 N.S.G. Hyslop, Trop. Anim. Health Prod., 4(1972)28.

42 J. Reiss, Z. Bakt., Abt. II, 130(1975)157.

43 W.R. Solomon, J. Allergy Clin. Immunol., 56(1975)235.

44 C. von Pirquet, Munchener Medizinische Wohchenschrift, 30(1906)1457.

45 J. Lacey, J. Pepys and T. Cross, in D. Shapton and R. Board (Editors), Safety in Microbiology. Society of Applied Bacteriology Technical Series No. 6, 1972, p. 151.

46 W.R. Solomon, in J.N. Pitts and R.L. Metcalf (Editors), Advances in Environmental Sciences, Vol. 1, John Wiley & Sons N.Y., 1969, p. 197.

47 D.C.F. Muir, in D.C.F. Muir (Editor), Clinical Aspects of Inhaled Particles, Heinemann, London, 1972, p. 1.

48 J. Pepys, Monographs in Allergy, Vol. 4, 1969, p. 21.

49 I.W.B. Grant, W. Blyth, V.E. Wardrop, K.M. Gordon, J.C.G. Pearson and A. Mair, Br. Med. J., 1(1972)530.

50 J. Lacey and M.E. Lacey, Trans. Br. Mycol. Soc., 47(1964)547.

51 J. Lacey, Ann. Appl. Biol., 69(1971)187.

52 E.F. Banaszk, W.H. Thiede and J.H. Fink, New Engl. J. Med., 283(1970) 271.

53 I.G. Popescu, E. Capetti, C. Galalaie and I. Spiegler, Rev. Raum. Med. Med. Interne., 13(1975)221.

54 P.H. Gregory, Proc. Roy. Soc. Lond. Ser. B., 177(1971)469.

55 D.A.J. Tyrrell, in P.H. Gregory and J.L. Monteith (Editors), Airborne Microbes, Seventeenth Symposium Soc. Gen. Microbiol., C.U.P., 1967, p. 247.

56 P.H. Gregory, The Microbiology of the Atmosphere, Interscience, New York, 1961, p. 33.

57 H.D. Covelli, J. Kleeman, J.E. Martin, W.L. Landau and R.L. Hughes, Amer. Rev. Respir. Dis., 108(1973)698.

58 M.R. Pereira and M.A. Benjaminson, Symp. Aerobiology and Man in the Americas, Amer. Acad. Adv. Sci., Mexico City, 1973.

59 P.H. Gregory, The Microbiology of the Atmosphere, John Wiley & Sons, New York, 2nd ed., 1973.

60 W.F. Wells, Amer. J. Publ. Health, 23(1933)58.

61 A.A. Andersen and M.R. Andersen, Appl. Microbiol., 10(1962)181.

62 A.A. Andersen, J. Bact., 76(1958)471.

63 K.R. May, Appl. Microbiol., 18(1969)513.

64 C.R. Albrecht, M.S. Thesis, Univ. of Florida, Gainesville, 1958.

65 H.W. Wolf, P. Skaliy, L.B. Hall, M.M. Harris, H.M. Decker, L.M. Buchanan and C.M. Dahlgren, Sampling Microbiological Aerosols, Public Health Monograph No. 60, Pub. Hlth. Service Publ. No. 686, U.S. Dept. of Health, Education and Welfare, 1959.

66 F.B. Higgins, Ph.D. Thesis, Georgia Inst. of Technol. Atlanta, 1964

67 C.W. Randall and J.O. Ledbetter, Amer. Ind. Hyg. Assn. J., 27(1966)506.

68 H.M. Decker and M.E. Wilson, Appl. Microbiol., 2(1954)267.

69 D.C. Blanchard and L.D. Syzdek, J. Geophys. Res., 77(1972)5087.

70 R.F. Hatcher and B.C. Parker, Va. J. Sci., 26(1975)141.

71 B.M. Smith, Ph.D. Thesis, Georgia Inst. Of Technol., Atlanta, 1968.

72 H.W. Horrocks, Proc. Roy. Soc. Lond. Ser. B., 79(1907)531.

73 G.M. Fair and W.F. Wells, Proc. 19th Annual Meeting New Jersey Sew. Works Assn., Trenton, N.J., 1934.

74 J.O. Ledbetter, Water and Sewage Works, 133(1966)43.

75 J.O. Ledbetter and C.W. Randall, Ind. Medicine and Surgery, 34(1965) 130.

76 P.J. Napolitano and D.R. Rowe, Water and Sewage Works, 113(1966)480.

77 F.C. Ladd, M.S. Thesis, Oklahoma State Univ., Stillwater, 1966.

78 R.C. Merz, Publ. No. 18, Univ. Southern California, Eng. Center, Los Angeles, Cal., 1957.

79 C.A. Sorber, S.A. Schaub and K.J. Guter, U.S. Army Medical Environ. Eng. Res. Unit Rept. No. 73-02, Army Medical Res. and Development Command, Edgewood Arsenal, Md., 1972.

80 C.A. Sorber, in J.F. Malina and B.P. Sagik (Editors), Virus Survival in Water and Wastewater Systems, University of Texas, Austin, 1974, p. 241.

81 K. Schultze, Bull. Hyg., 19(1944)372.

82 H. Reploh and M. Handloser, Arch. Hyg. Berl., 141(1957)632.

83 E.M. Shtarkas and D.G. Krasil'shchikov, Abst. Hyg. and San., 35(1970) 330.

84 E.D. King, R.A. Mill and C.H. Lawrence, J. Environ. Hlth., 36(1973)50.

85 A. Woratz, Gesundh Wes. Desinfekt, 55(1963)145.

86 Final Report of Committee on Standard Methods for the Examination of Air, Amer. J. Pub. Hlth., 7(1917)154.

87 M.C. Henry, J. Findlay, J. Spangler and R. Ehrlich, Arch. Environ. Health, 18(1969)580.

88 E. Sepp, J. Environ. Eng. Div. Proc. Amer. Soc. Civil Eng., 99(1973)123.

89 K.E. Jensen, Bull. World Health Org., 10(1954)171.

90 R.M. Buchan, R.A. Mill and C.H. Lawrence, J. Environ. Hlth., 35(1973) 342.

91 N.A. Clarke and P.W. Kabler, Health Lab. Sci., 1(1964)44.

92 K.F. Fannin, J.C. Spendlove, K.W. Cochran and J.J. Gannon, Appl. Environ. Microbiol., 31(1976)705.

93 H. Finkelstein, Preliminary Air Pollution Survey of Biological Aerosols - A Literature Survey, Publ. No. APTD 69-30, Natl. Air Poll. Control Admin., 1969.

94 K. Popek, E. Kopečna, N. Bieronska, Z. Cerny, B. Janicek and Z. Kozusnik, Zentr. Bakteriol. Parasitenk. Abt. I Orig., 210(1969)502.

95 R.W.S. Harvey, T.H. Price and D.H.M. Joynson, J. Hyg., 76(1976)91.

96 H.F. Dawling, Bact. Rev., 30(1966)485.

HEALTH HAZARDS ARISING FROM WATER-BORNE PATHOGENS

The release of faecal pollution into water can introduce a wide variety of intestinal pathogens. There are outbreaks of gastro-intestinal illness of unknown aetiology which are characterized by diarrhoea, abdominal cramps and, occasionally, vomiting and fever. These can arise from drinking sewage contaminated water. This type of illness was responsible for 54% of the outbreaks and 62% of the cases of water-borne disease in the United States in 1973 [1]. The outbreaks tended to show a seasonal variation with over 60% of cases occurring in the months of June, July and August when use of recreational areas is most intense. The cases can occur following the use of untreated groundwater, and in 1973 nearly 200 cases in the U.S.A. resulted from a cross-connection between a water main and a pressure sewer main.

The organisms which are well characterized and most commonly present in sewage polluted water can include strains of Salmonella, Shigella, Leptospira, enteropathogenic Escherichia coli, Francisella, Vibrio, Mycobacterium, human enteric viruses, cysts of Entamoeba histolytica or other pathogenic protozoans, and larvae of various pathogenic worms (Table 1).

Bradford Hill [2] listed nine criteria which must be fulfilled when judging whether an environmental pollutant (which includes micro-organisms) is causing a disease. They are designed to show that there is a definite association between a polluted environment and a communicable disease.

A cause and effect situation has been established in the case of many microbial diseases. Salmonella strains have frequently been detected in sewage, streams, irrigation waters, wells and tidal waters, and this has been correlated with outbreaks of salmonellosis. Other pathogenic organisms are isolated from water less frequently, which may be due to greater difficulties being encountered in their isolation and identification. Moore [3] showed that the monitoring of sewage was a useful epidemiological tool in determining which diseases are present in a community. Human pathogens can also be found in the gut of other warm blooded animals and these organisms are acquired from the consumption of contaminated food or water [4]. The faecal pollution arising from any source is therefore important.

It must always be borne in mind that microbial diseases are highly complex host-parasite interactions. There is not necessarily a simple

relationship between the presence of a pathogenic organism and the relevant disease in man. The problem of assessing the importance of a pathogen in a polluted environment was shown by Perry et al. [5], who demonstrated that when haemolytic streptococci were released into the environment they rapidly lost their virulence, and so had no relevance in the
epidemiology of streptococcal infections. Brison [6] suggested that
Staphylococcus aureus was possibly a hazard from seawater, but this
organism is present in the upper respiratory tract of over 25% of the
population, so in this case infection depends on individual susceptibility
not just on exposure to the organism.

Bacteria

i)   Salmonellosis

This is seen in humans as an acute gastroenteritis with diarrhoea and
abdominal cramps. Fever, nausea, vomiting and headache are often seen as
symptoms. Recent U.S.A. usage recognized only three species, namely
S. choleraesuis, S. typhi and S. enteritidis. The first two do not contain
subtypes (serotypes) whilst the third has over 1,000 serotypes. However,
the classical names still predominate in the literature and show little
sign of being displaced. The aetiological agent for typhoid fever is
Salmonella typhi which is specific for man and is not seen in other
animals.

The incidence of salmonellosis in humans is low and shows a seasonal
variation. About 10^5 to 10^7 organisms have to be ingested to give rise to
the disease.

Weibel et al. [7] estimated that in the U.S.A. between 1946 and 1960,
only 1.4% of the total typhoid morbidity was due to water-borne infection,
compared to an estimated 40% in 1908. However, one must not be too com-
placent about the situation, even in the developed countries, because in
1973 [1] there was an outbreak of typhoid in Dade Co., U.S.A., in which
there were 225 cases due to using contaminated well water. This was the
largest outbreak in the U.S.A. since 1939.

The average number of persons excreting Salmonella at any time is not
definitely known, but Hall and Hauser [8] and Public Health Laboratory
Service [9] suggests that it is less than 1% of the population in the
U.S.A. and United Kingdom, whilst Gadasekharam and Velaudapillai [10] state
that it is 3.9% in Ceylon. From these observations negative results would
be obtained for the presence of Salmonella in domestic sewage, especially

(Continued on p. 56)

Table 1.

ORGANISMS WHICH CAN BE PATHOGENIC TO MAN AND ARE TRANSMITTED

BY CONTAMINATED WATER

Organism	Disease	Principal Site which is Affected
1. BACTERIA		
*Salmonella typhi	Typhoid fever	
*Salmonella choleraesuis) *Salmonella enteritidis and) other serotypes)	Enteric fevers Gastroenteritis	Gastro-intestinal tract
*Shigella sp.	Dysentery	Gastro-intestinal tract
*Vibrio cholerae	Cholera	Lower intestine
*Enteropathogenic Escherichia coli	Gastroenteritis	Gastro-intestinal tract
Francisella tularensis	Tularaemia	Respiratory tract, gastro-intestinal tract, lymph nodes
Leptospira icterohaemorrhagiae	Leptospirosis	Generalized
Mycobacterium tuberculosis	Tuberculosis	Lungs and other organs
2. PROTOZOA		
*Entamoeba histolytica	Amoebiasis	Gastro-intestinal tract
Giardia lamblia	Giardiasis	Gastro-intestinal tract
Naegleria gruberi	Amoebic meningoencephalitis	Central nervous system

Table 1.(continued)

Organism	Disease	Principal Site which is Affected
3. PARASITIC WORMS		
*Taenia saginata	Beef tapeworm	Gastro-intestinal tract
*Ascaris lumbricoides	Ascariasis	Small intestine
*Schistosoma mansoni) *Schistosoma japonica) *Schistosoma haematobuim)	Schistosomiasis	Bladder
Necator americanus) Ancylostoma duodenale)	Ancylostomiasis	Gastro-intestinal tract
Diphyllobothruim latum	Diphyllobothriasis (Fish tapeworm)	Gastro-intestinal tract
Echinococcus granulosus	Echinococcosis	Liver and lungs
Anisakis sp.	Anisakiasis	Gastro-intestinal tract

* More commonly occurring infections.

from a small population. Therefore, any test which gives such a negative result for Salmonella cannot be taken to confirm the absence of this or other water-borne pathogens. Salmonellosis may not be the most serious water-borne disease, but the organisms are probably easier to isolate from water, food and faeces than many of the other pathogenic organisms and the numbers are readily correlated with the level of E. coli.

There are a large number of serotypes of Salmonella which are pathogenic to man and the frequency of their isolation can vary from country to country and from year to year. For example, in 1965 the 10 most common Salmonella serotypes which were isolated from human sources were S. typhimurium, S. heidelberg, S. newport, S. infantis, S. enteritidis, S. saint-paul, S. typhi, S. derby, S. oranienberg and S. thompson [11]. Grunet and Nielsen [12] reported that in Denmark for 1960-1968 the corresponding serotypes were S. typhimurium, S. paratyphi B, S. enteritidis, S. newport, S. typhi, S. infantis, S. indiana, S. montevideo, S. blockley and S. muenchen.

Salmonellae can be isolated from perfectly healthy farm animals, 13.4% of pigs [13] in England and Wales, 13% of cattle in the U.S.A. [14] and up to 15% of sheep are symptomless carriers. Andre et al. [15] showed that salmonellae will survive for fourteen to sixteen days at 21°-29°C in farm pond water, whilst Pollach [17] has shown that cattle rearing, cattle markets and abattoirs can all produce effluent that will pollute water with salmonellae, and this contamination can finish up in streams and lakes which may be used for recreational purposes or for drinking. The serotypes which can be isolated from diseased and healthy cattle include S. typhimurium, S. derby, S. dublin, S. oranienburg, S. java, S. choleraesuis, S. anatum, S. infantis, S. abony, S. newington, S. stanley, S. meleangridis and S. chester [16,17,18]. Epidemics have been produced in humans by at least six of these thirteen serotypes.

Over 620 strains belonging to six serological groups and 34 sero- and lysotypes of salmonellae were isolated from a sewage outfall in the Danube at Bratislava [18]. The most common were S. paratyphi B, S. bareilly, S. typhimurium and S. anatum.

Salmonellae have been detected in many polluted water systems including sewage, stabilization ponds, water used for irrigation, streams, estuaries and tidal waters [19].

The survival of salmonellae in water has been studied by several workers and is found to follow the same pattern as the organisms used as indicators of pollution. If the water is polluted with nutrient-rich waste, has a low stream temperature and then receives an inoculum of salmonellae the disappearance of the organisms can be significantly delayed. The U.S. Department of Health, Education and Welfare, Public Health Service [11] examined the water of the Red River, N. Dakota, U.S.A. and showed that salmonellae could be isolated 22 miles downstream from the sewage discharge of Fargo, N. Dakota and Moorhead, Minnesota during September. Because sugar beet processing waste was discharged under the ice in November, salmonellae were isolated 62 miles downstream. In January, they could be detected 73 miles downstream, which is four days flow time from the source of pollution.

ii) Shigellosis

Hardy and Watt [20] gave exposure to Shigella as the most commonly identified cause of diarrhoea in the U.S.A. Also Hughes et al. [1] stated that shigellosis was the illness of known aetiology which was responsible for 17% of the outbreaks and 19% of the cases of water-borne disease in U.S.A. in 1973. The disease in epidemics is mainly spread by person to person contact or through eating contaminated food. However, there has been a significant increase in the number of outbreaks arising from drinking poor quality water. These cases may result from breaks occurring in water treatment systems [21] or where water supplies receive no treatment at all [22]. Breaks in sewer pipes and sewage seeping into water supply pipes have also been implicated [23] and so has flood water carrying excreta into well supplies [24]. There was also an explosive outbreak of shigellosis in a cruise ship in 1973 which affected 690 passengers and crew, and was traced to sewage contamination of the drinking water coupled with improper chlorination procedures.

It has been estimated that the average number of people excreting Shigella in the population is 0.46% in the U.S.A. [25], 0.33% in England and Wales [9] and 2.4% in rural areas of Ceylon [10]. There are at least 32 Shigella serotypes with S. sonnei and S. flexneri accounting for over 90% of the isolates. The shigellae are rarely found in animals other than man and are less invasive than salmonellae.

Once Shigella sp. leave the gut their survival is governed by the environmental conditions. Bartos et al. [26] showed that Shigella could be recovered up to twenty-two days after being introduced into a well. The work of Dolivo-Dobroval'skii and Rossovaskaia [27] demonstrated that

Shigella could survive in river water for up to four days, but if the
water was aerated then the survival time was reduced to 30 min. Nakamura
et al. [28] showed that Shigella can survive for several days in seawater
and will therefore tolerate estuarine conditions.

The problem with studying the occurrence of Shigella in natural bodies
of water, as well as under polluted conditions, lies in the complicated
and time consuming techniques required for identification. These can only
be carried out by highly trained technical staff. Also, the picture may
be complicated further by the observation that the colonial characteristics of Shigella may change after a time in a natural environment [15].

iii) Enteropathogenic E. coli

There are 14 different distinct serotypes of E. coli which may cause
gastroenteritis, which is seen as a profuse watery diarrhoea with little
mucous and no blood. There is generally nausea and dehydration but
usually no fever. The serious diarrhoea seen in children, under five
years of age (particularly the newborn) is often caused by enteropathogenic
E. coli [29]. The disease may also affect adults and may have a role in
"travellers' diarrhoea" and food poisoning incidents.

The carrier state in the general population is not easily determined,
but an indication may be obtained. Hall and Hauser [8] reported a level
of 6.4% in food handlers in Louisiana, Public Health Laboratory Service a
level of 2.4% in children in England and Wales, and Rottini and Zacchi
[29] found 3.3% healthy newborn children were carriers in Trieste.

When nearly 2,000 water samples were examined in Europe, 21.5% were
found to be microbiologically unsafe with a high number of E. coli
present, 2.3% of these being from enteropathogenic groups. The number of
viable enteropathogenic E. coli was 8% of the total number of viable
E. coli present [30].

The gastroenteritis seen in adults probably comes from drinking
water contaminated with enteropathogenic E. coli, and outbreaks from such
a source have been reported from Washington [31], Uppsala [32], Gimo,
Sweden [33] and northern France [34], and in many cases the organisms were
removed by adequate chlorination. Lanyi et al. [35] reported that an outbreak of gastroenteritis in a children's camp in Hungary was due to
E. coli serotype 0124:K72, H:32 which was entering a spring (used for
drinking water) from a leaking sewer.

If enteropathogenic E. coli are present in sewage polluted lakes or

streams they probably represent less than 1% of the coliform population.
The survival of E. coli depends on the usual environmental factors
including temperature, pH, nutrient level, deposition with sediment or
presence of predators. Usually multiplication is suppressed by dilution
and in two to five days the level of viable organisms can be reduced to
10% of the original inoculum [36].

iv) Cholera

Cholera probably originated in the Far East, and has been endemic in
India for centuries. In the nineteenth century it spread to other parts
of the world, giving rise to pandemics in Europe. The development of
controlled water supplies which were separated from the sewage disposal
system dramatically reduced the incidence of the disease in Europe. In
recent years there has been a spread of cholera which could be due to lack
of quarantine enforcement combined with inadequate sanitary facilities in
some countries. The increased speed of the air transport of carriers or
of infected food imports may also be contributory factors.

The disease is caused by the bacterium Vibrio cholerae, which is seen
as three antigenic types Ogawa, Inaba and Hikojima. It is a serious
intestinal disease characterized by sudden diarrhoea with copious watery
faeces, vomiting, suppression of urine, very rapid dehydration, lowered
temperature and blood pressure and complete collapse. The death of the
patient can occur within a few hours and the disease has a 60% mortality
without adequate therapy, but this can be reduced to less than 1% if fluid
and electrolytes are replaced by intravenous transfusion. The disease
presents no problem in the developed countries.

Most cholera vibrios are not haemolytic, but a strain which forms a
soluble haemolysin was first isolated in 1906 at the El Tor quarantine
station in the Gulf of Suez. These and subsequently isolated strains have
the same pathogenic properties as V. cholerae but they are considered to
be a separate biotype. The El Tor vibrios are more resistant to chemical
agents, survive longer outside the human body and persist for longer in
man than the classical strain.

Healthy symptomless carriers of V. cholerae may be from 1.9% to 9.0%
of the population [37], while the El Tor vibrios may be present in 9.5%
to 25% [38]. Cholera can be spread by person to person contact, although
this is not easily done as 10^8 to 10^9 organisms are required to cause the
illness. More commonly it is transmitted by eating food contaminated

after it has been washed with polluted water, or food handled by a carrier, or by drinking polluted water.

Pesigan [39] showed that V. cholerae can survive for from one hour to thirteen days, surviving better in the absence of other organisms or organic matter and at a pH of 8.2 to 8.7. Because the vibrios may survive for only a short time in a heavily polluted environment, the continued addition of faeces from infected persons is probably required to keep the viable numbers of organisms at an infective level.

Sen and Jacobs [40] have shown that chlorination with levels of 2.0 to 3.0 mg per litre for 10 min contact is insufficient to produce potable water (free of V. cholerae or Salmonella) from the polluted water of the Hoogly River, India. The organisms were probably protected by clumping, from the effects of the chlorine during the disinfection period.

v)   Tularaemia

The causal organism in this disease is Francisella tularensis (also called Pasteurella tularensis and Bacterium tularense). The pathogen usually enters through skin abrasions or mucous membranes to give chills and fever, swollen lymph nodes and a general malaise. Without treatment the illness is usually protracted with delirium and coma, and may be fatal. An ulcer may develop at the site of the initial contact and the organism can be found in the lesions up to a month from the start of the disease. It is not considered to be transmitted from person to person.

Human cases are most often caused by contact with infected wild animals or contact with wood ticks [41]. The disease can be spread by drinking water contaminated with the urine, faeces or cadavers of several species of rodents or rabbits [42]. Outbreaks of tularaemia following drinking of contaminated water have been reported from the U.S.A. [43,44], in the U.S.S.R. [42] and Czechoslovakia [45]. Also, Schmidt [46] reported large outbreaks of tularaemia in the Soviet army, when in the Rostov region during the Second World War. In this instance contaminated drinking water was implicated.

It is difficult to ascertain the viable numbers of F. tularensis in water because of the problem of recovering the organisms. Parker et al. [47] were able to show that stream water was contaminated only by using guinea pig inoculation, and all attempts at isolation on cysteine-heart infusion agar plates were unsuccessful.

Francisella tularensis, like most of the water-borne pathogens, will

survive better in cold water and the presence of nutrients may extend its
survival.

vi) Leptospirosis

This disease is caused by a group of motile bacteria characterized
by being formed from very fine spirals which are wound so tightly that
they are barely distinguishable under the microscope. The Leptospira
infect the bloodstream after entering through abrasions or mucous membranes.
The infection may eventually involve the kidneys, liver and central
nervous system.

Many serotypes of Leptospira have been isolated, and at least 18
serological groups and 100 serotypes are recognized. The first human
leptospiral infection to be recognized was the severe febrile illness,
Weil's disease, which was shown in 1915 to be caused by L. icterohaemorrhagiae transmitted to man by infected rats. All common species of
Leptospira are pathogenic to man, except saprophytic L. biflexia which
inhabits small streams, lakes or stagnant water.

The disease may be found in the Americas, Far East, Middle East and
Europe. The level of infection in the population is probably less than
1% and Torten et al. [48] reported that it was 0.37% in Israel. The level
can be much higher, up to 3%, when persons involved in meat processing,
sewage disposal, or handling polluted water e.g. in irrigation work are
considered.

The disease is transmissible from animals to man, and in the case of
water-borne infection the organism enters rivers, streams etc. through the
direct urination of cattle, pigs or wild animals. Pathogenic leptospires
were found in 43% of effluent samples from large piggeries and the organisms survived for a maximum of 96 h [49]. This may present a hazard when
pig effluent is used as a fertilizer. Contamination of water with animal
effluent has led to outbreaks of leptospirosis in Holland, for instance,
where the canals have been used for swimming [50].

Leptospires survived for a short time when mixed with faecal material
and the system was kept at either 5° or 37°C [51], but survival was much
higher in sterile tap water at 25°-27°C and viable organisms could be
recovered up to thirty-seven days after inoculation [52]. If a mixed
bacterial culture was added to this latter system, then the survival of
L. icterohaemorrhagiae was reduced by 50%. These findings suggest that
the conditions under which any effluent is kept can have a significant

effect on the viable numbers of bacteria subsequently found.

vii) Tuberculosis

Pulmonary tuberculosis is caused by Mycobacterium tuberculosis and the closely related M. bovis, and can be transmitted by water, but this mechanism is relatively uncommon. It may take many years before the disease is seen in a clinical form which makes it extremely difficult to trace a particular infection back to a contaminated water source. The mycobacteria pathogenic to man, which are associated with water include M. tuberculosis, M. balnei (marinum) and M. bovis.

Laird et al. [53] showed that although organisms are found normally in the sputum of infected persons, they can also be isolated from their faeces. Cases of tuberculosis in children followed a near drowning incident in water contaminated with sewage [54]. Also, the use of a warm spring water pool for swimming in Colorado gave rise to 124 cases of granuloma and 1,000 organisms/ml of M. balnei were recovered from the water.

Mycobacteria can be isolated from swimming pool water, although they are usually the non-pathogenic M. aquae which can resist 2.0-2.5 mg/l. of free chlorine [55]. Saprophytic mycobacteria can also be isolated from potable water supplies [55].

The effluent discharged from farms, abattoirs or cattle markets could also ultimately contaminate drinking water supplies. Mycobacterium tuberculosis can survive in aquatic environments for several weeks because Rhines [56] demonstrated the survival of the avian strain for seventy-three days in raw sewage or estuarine water. However, the organism may lose its virulence during this time.

The atypical mycobacteria M. kansasii and M. xenopi which can cause skin lesions and pulmonary tuberculosis, have been found in water systems in England [57,58], and in tap water in California [59].

Goslee and Wolinsky [60] studied the mycobacterial flora of 321 water samples, including natural waters, water treated for drinking purposes and water in contact with animals. The highest yield of positive cultures came from samples in contact with zoo animals and fish. The isolated strains were mainly slow growing mycobacteria particularly M. gordonae and from the M. avium-intracellulare-scrofulaceum complex. They concluded that water may be contaminated with potentially pathogenic mycobacteria and so serve as a source for human infection.

Viruses

Theoretically, any human virus which is excreted in faeces could be transmitted through contaminated water supplies. The viruses which are most likely to be transmitted by such a route are infectious hepatitis, the enteroviruses (polio virus, echovirus and coxsackie virus), reovirus and adenovirus. In many cases the involvement of a water-borne route is inconclusive.

In recent years, infectious hepatitis has been established as a water-borne disease. Moseley [61] accepted that in at least 30 of the 50 reported outbreaks of the disease the evidence was sufficient to implicate a water-borne virus. In all instances there had been a large and sustained contamination of the water supply with sewage.

Viral hepatitis is now subdivided into three forms: hepatitis A (epidemic, infectious or short-incubation-period hepatitis), hepatitis B (homologous serum or post-transfusion hepatitis) and a third form for which the term hepatitis C ("non-A-non-B" hepatitis) has been proposed. Hepatitis A can be transmitted by water and can occur endemically or epidemically. Hepatitis B was believed to spread only by inoculation, but it is generally accepted that it could be spread by close personal contact and it is highly endemic in certain countries e.g. Greece [62].

Transmission of hepatitis A is similar to that of the water-borne diseases caused by enteric bacteria. From a public health point of view, faecal contamination of food, drinking water and swimming pools or lakes is the most likely cause of epidemics. The virus is excreted in the faeces during the early acute phase of the disease and perhaps during the incubation period. Chronic carriers for this disease have not been isolated, although they probably exist.

The largest water-borne infectious hepatitis epidemic occurred in Delhi during 1955-56 and involved some 28,000 cases, with possibly ten times this number remaining subclinical [63]. When the water supply of Delhi was most polluted, almost all the water intake from a channel of the Jamina River was drawn from a canal which was nearly an open sewer. This water was receiving the excreta from a large number of refugees, who were already suffering from an outbreak of infectious hepatitis. Although the water supply received chlorination, the water was turbid and the solid material could have protected the virus. During this epidemic there was no parallel increase in infections due to water-borne pathogenic bacteria, which suggests that the chlorination treatment which is sufficient to eliminate bacteria may be insufficient to destroy viruses.

Outbreaks of pharyngo-conjunctival fever have been traced to swimming pools [64], and in one case infections caused by adenovirus 3 were caused by swimming pool water which had not been chlorinated for 36 h over a week-end [65]. Once the chlorination system was working, there were no further cases.

There has been increased interest in filterable agents which can cause gastroenteritis. These have not been reliably cultivated in the laboratory. Weibel et al. [66] showed that about two-thirds of the water-borne outbreaks of gastroenteritis described in the literature were not due to any known pathogen. Lobel et al. [67] examined an outbreak of gastro-enteritis in visitors to Pennsylvania State Park which was traced to the park water supply and could have been viral in origin. More recently Dolin et al. [68] produced further evidence for a possible viral aetiology of gastroenteritis, and Gomez-Barreto et al. [69] have suggested the involvement of reovirus-like agents as a possible cause of one outbreak of this condition. However, the cause and effect have not been definitely shown, and Koch's postulates have yet to be fulfilled.

There is some evidence that warm blooded animals may carry viruses which can infect humans [70]. Lundgren et al. [71] found human enteric viruses in about 10% of beagles. If such animals can act as reservoirs of infective agents, then any pollution (not just with human faeces) of water, e.g. streams, rivers or storm water may constitute a potential hazard.

The survival of human enteric viruses is governed by the same factors as the survival of bacterial pathogens; water temperature and the presence of bacteria seem to be important. Clarke et al. [72] found that Coxsackie virus A2 survives better in distilled water than in sewage, and better in seawater which has been heated for 1 h at 45°C [73]. Also coxsackie virus can be recovered from sewage after forty days at 20°C and sixty-one days at 8°C. However, Lundgren et al. [74] showed that in samples from the Ohio River, no viruses were found after six days at 20°C or sixteen days at 8°C. Polio virus, coxsackie virus and echovirus survive at 20°C, and at 4°C will survive for up to nine weeks in pond water [75]. If magnesium chloride is added then the survival time is extended up to twelve weeks, but if iron salts are added then it is reduced to less than three weeks. The level of viruses is reduced by dilution in running water and by adsorption onto clay particles followed by deposition.

Protozoa

Amoebiasis is a disease of the large intestine which may be seen as mild abdominal discomfort with diarrhoea alternating with constipation, or it may be a chronic dysentery with a discharge of mucous and blood. The causal agent is a parasitic protozoan, Entamoeba histolytica which is transmitted mainly through drinking water which has been contaminated with faeces, although contaminated food can be implicated in certain circumstances.

The carrier state is very important and the level has been estimated at 10% in Europe, 12% in the Americas (with 3-5% in the U.S.A. [76]), 16% in Asia and 17% in Africa. However, there is correlation between the socio-economic group and the incidence of the carrier state. This is due to poor personal hygiene, unprotected water supplies and inadequate waste disposal. Also, the occupation may be an additional hazard, for 9% of agricultural workers in irrigated fields and 14% of sewage plant operators were found to be carrying E. histolytica. Le Maistre et al. [77] showed that E. histolytica could get into water supplies from many sources including cross connection sewers and water supply pipes, back siphoning from toilets, drainage from defective sewer pipes over open water and from leaking water pipes with low pressure that were submerged in sewage. If attention is paid to using correct plumbing practices and preventing cross connection, this will prevent most of the contamination occurring.

The level of E. histolytica in sewage is usually very low, for example, at Haifa, Israel, Kott et al. [78] found, in raw sewage, 5 cysts/l and in the effluent from a sewage treatment plant 1-2 cysts/l. The cysts will survive for more than 150 days in good quality water at $12^{\circ}-22^{\circ}C$ [79], and in sewage or natural bodies of water the cyst persistence is reduced by 30% for each $10^{\circ}C$ rise in water temperature [80]. Thus the cysts can remain viable in sewage or water supplies for a long time, but this is made less important by the relatively low initial levels. There is also a reduction in numbers when the effluent is discharged into streams, by dilution and settling out.

Giardia lamblia, which causes giardiasis, is an intestinal protozoan which, most commonly, gives rise to a protracted intermittent diarrhoea. The organism is found world-wide and the carrier rate ranges from 1.5% to 20% in different parts of the U.S.A., and again there is a dependence on the socio-economic and age of the group under examination. Moore et al. [81] reported an epidemic of giardiasis in 1965 to 1966 at Aspen, Colorado which affected 11.3% of 1,094 skiers. The cause was contamination of well water (used for drinking purposes) by sewage leaking from defective pipes, which passed near the wells. Cysts of G. lamblia were found in the sewage and in the faeces of 6.9% of the permanent residents using the defective sewage system. An outbreak of giardiasis in Colorado was traced to contaminated groundwater [1], and another in Leningrad to tapwater [82].

Naegleria species are free living amoebae common in soil, sewage effluents and surface waters. However, pathogenic strains of Naegleria gruberi cause the fatal disease amoebic meningocephalitis. The organism can enter the body via the upper respiratory tract and rapidly migrates to the brain, cerebrospinal fluid and the blood stream [83]. The symptoms which are characteristic of the disease appear four to seven days after water contact and death follows four to five days later. Amoebic meningocephalitis has been associated with swimming and diving in lakes [83], an indoor swimming pool [84] and a polluted estuary [85]. Epidemics have been reported from Florida, South Australia and Czechoslovakia.

The cysts of N. gruberi and N. fowleri are sensitive to chlorination [86].

Parasitic Worms

Many parasitic worms are found in sewage and these can present a hazard, particularly to sewage plant workers. Problems may also arise from lakes used for recreational purposes which have been contaminated either with sewage or from farmyards. Farmworkers using sewage contaminated water for irrigation are also at risk. There may be a risk of infection following drinking of contaminated water, although modern methods of treatment should eliminate this problem. This may be more of a danger where water is re-used.

The eggs of Taenia saginata (beef tapeworm) have been found in sewage at a rate of up to 2 eggs/100 ml [87]. The eggs are very resistant and can survive for over 330 days in cool, moist conditions which are similar to those of liquid sewage sludge [88]. Dried sewage sludge may be considered to be free from viable eggs after one year's storage before agricultural use.

Ascariasis is an intestinal infection most frequently found in young children. The large intestinal round worm, Ascaris lumbricoides, is the infecting agent and embryonated eggs are transmitted through faecal contamination in water, soil and on vegetables. About 200,000 eggs can be produced each day and these can remain viable for several months in the soil, provided suitable conditions exist. Sinnecker [89] found A. lumbricoides eggs in 2% of faeces from sewage plant operators, and in 16% of faeces from farm workers (using irrigation with sewage for crop cultivation). Baumhogger [90] showed that raw sewage in Darmstadt, Germany contained 540 Ascaris eggs/100 ml. This high level was correlated with an outbreak of ascariasis which affected 90% of the population. A level of 0.11 organisms/100 ml of river water was reported by Wang et al. [91] following the discharge of chlorinated primary sewage effluent.

Schistosomiasis is a debilitating disease in which the adult male and female worms live in the veins of humans. The migration of the blood fluke causes complications in chronic infections. The infection can be acquired by swimming or wading through freshwater (during rice cultivation) which is infected with the larval stage (cercariae) of Schistosoma mansoni, S. japonicum or S. haematobuim, the latter being the aetological agent of schistosomal haematuria and schistosomal carcinoma of the bladder.

The life cycle of the parasite depends on one phase in an appropriate snail and another phase in man, rodents or farm animals. The continuation of the life cycle is dependent on urine and faecal material from the infected animal reaching streams and lakes, where the eggs hatch and the larval stage (miraciduim) finds a suitable freshwater snail as host. If the snail population can be reduced there will be a reduction in the incidence of the disease although control of the sanitation facilities is a more effective and practical measure. For example, the transmission of S. mansoni has been controlled to a large extent in St. Lucia by the provision of adequate domestic water supplies [92]. The disease can be found in Africa, Middle East, Far East, South America and the Caribbean islands.

There are several other diseases transmitted by faecal material which contaminates water supplies, including hookworm disease (ancylostomiasis) which is associated with the helminths Necator americanus and Ancylostoma duodenale. In the Soviet Union [93] there was an outbreak of fish tape-worm disease (diphyllobothriasis) following discharge of untreated domestic sewage into a reservoir. This disease follows eating raw or inadequately cooked fish which are the intermediate hosts for Diphyllobothruim latum. Occasionally humans can be infected with dog tape-worm larvae (Echinococcus granulosus) with the development of cysts chiefly in the liver and lungs. The disease usually comes from close contact with infected dogs, but in a few cases the cause was thought to be water supplies which were contaminated by the faeces of dogs or wolves.

The disease anisakiasis affects the gastro-intestinal tract and is caused by eating raw or inadequately cooked herring and other salt water fish which contain the larval stage of a nematode. The infective agent is <u>Anisakis</u> sp., but the complete life cycle is not understood.

REFERENCES

1   J.M. Hughes, M.H. Merson, G.F. Craun and L.J. McCabe, J. Infect. Dis., 132(1975)336.
2   A. Bradford Hill, Proc. R. Soc. Med., 58(1965)95.
3   B. Moore, Monthly Bull. Ministry of Health, 7(1948)241.
4   J.L. Summers, The Sanitary Significance of Pollution of Waters by Domestic and Wild Animals. A literature review. U.S. Dept. of Health, Education and Welfare, Public Health Service, Shellfish Sanitation Technical Report, 1967.
5   W.D. Perry, A.C. Siegel and C.H. Rammelkamp, Am. J. Hyg., 66(1957)96.
6   J. Brison, Bull. Wld. Health Org., 38(1968)79.
7   S.R. Weibel, F.R. Dixon, R.B. Weidner and L.J. McCabe, J. Amer. Water Works Assoc., 56(1964)947.
8   H.E. Hall and G.H. Hauser, Appl. Microbiol., 14(1966)928.
9   Public Health Laboratory Service and Society of Medical Officers of Health, Monthly Bull. Ministry of Health and Public Health Lab. Service, 24(1965)376.
10 J. Gadasekharam and J. Velaudapillai, Z. Hyg., 147(1961)347.
11 U.S. Department of Health, Education and Welfare, Public Health Service, Report on Pollution of Interstate Waters of the Red River of the North (Minnesota, North Dakota). Robert A. Taft Sanitary Engineering Center, Ohio, 1966.
12 K. Grunnet and B.B. Nielsen, Appl. Microbiol., 18(1969)985.
13 E. Prost and H. Riemann, Ann. Rev. Microbiol., 21(1967)495.
14 H.J. Rothenbacker, J. Amer. Vet. Med. Assoc., 147(1965)1211.
15 P.M. Nottingham and A.J. Urselmann, N.Z. J. Agr. Res., 4(1961)449.
16 W. Pollach, Vien Tierargth Mschr., 15(1964)161.
17 J.R. Miner, L.R. Fina and C. Piatt, Appl. Microbiol., 15(1967)627.

18 O. Kadlecova, Z. Bakt. Abt. II, 130(1975)144.

19 D.A. Andre, H.H. Weiser and G.W. Maloney, J. Amer. Water Works Assoc., 59(1967)503.

20 A.V. Hardy and J. Watt, Public Health Rep., 60(1945)57.

21 D.M. Green, S.S. Scott, D.A.E. Mowat, E.J.M. Shearer and J.M. Thomson, J. Hyg., 66(1968)383.

22 R.H. Drachman, F.J. Payne, A.A. Jenkins, D.C. Mackel, N.J. Petersen, J.R. Boring, F.E. Garean, R.S. Fraser and G.G. Myers, Amer. J. Hyg., 72(1960)321.

23 T. Nikodemusg and L. Ormay, Arch. Inst. Pasteur Tunis, 36(1959)43.

24 A.J. Rors and E.H. Gillespie, Monthly Bull. Min. Health, 11(1952)36.

25 L.B. Reller, E.J. Gangarosa and P.S. Brachman, Amer. J. Epidemiol., 91(1970)161.

26 D. Bartos, J. Bansagi and K. Bakos, Z. Hyg. Infekt., 127(1947)347.

27 L.B. Dolivo-Dobroval'skii and V.S. Rossovaskaia, Gigiena i Sanit., 21(1956)52.

28 M. Nakamura, R.L. Stone, J.E. Krubsack and F.P. Pauls, Nature, 203 (1964)213.

29 G.D. Rottini and T. Zacchi, Giarn Microbiol., 16(1968)189.

30 A.Z. Dragus and M. Tratnik, Z. Bakt. Hyg., I. Abt. Orig. B., 160(1975) 60.

31 S.A. Schroeder, S.A. Caldwell, J.R. Vernon, T.M. White, P.C. Granger and S.J. Bennett, Lancet, (i)(1968)737.

32 S. Bengtsson, R. Berg, D. Danielsson, K.M. Lundmark, F. Nordbring and O. Sandler, Lakartidningen, 63(1966)4499.

33 D. Danielsson, G. Laurell, F. Nordbring and O. Sandler, Acta Pathol. Microbiol. Scand., 72(1968)118.

34 P. Monnet, Ann. Inst. Pasteur, 87(1954)347.

35 B. Lanyi, J. Szita, B. Ringelhann and K. Kovach, Acta Microbiol. Acad. Sci. Hung., 6(1959)77.

36 R. Mitchell, Water Res., 2(1968)535.

37 R. Pollitzer, W.H.O. Monograph Series No. 43, World Hlth. Org., Geneva, 1959.

38 C.H. Yen, Cholera in Asia, in Proceedings of the Cholera Research Symposium, P.H.S. Pub. 1328(1965)346.

39 T.P. Pesigan, in Proceedings of the Cholera Research Symposium, P.H.S. Pub., 1328(1965)317.

40 R. Sen and B. Jacobs, Indian J. Med. Res., 57(1969)1220.

41 M.R. Bow and J.H. Brown, Am. J. Public Health, 36(1946)494.

42 M.I. Tsareva, J. Microbiol. Epidemiol. Immunobiol., 30(1959)34.

43 W.L. Jellison, D.C. Epler, E. Kuhrs and G.M. Kohls, Public Health
    Report, 65(1950)1219.

44 Communicable Disease Center Technology Branch, Summary of Investi-
    gations, 14(1958)April-Sept.

45 T. Mittermayer, A. Paurova, E. Belicova, A. Nickyova and H. Smilnicka,
    Cas Lek Cesk, 115(1976)354.

46 B. Schmidt, Z. Hyg. Infektionskrankh, 127(1947)139.

47 R.R. Parker, E.A. Steinhaus, G.M. Kohls and W.L. Jellison, Natl. Inst.
    Hlth. Bull. No. 193, U.S. Publ. Hlth. Service, 1951.

48 M. Torten, S. Birnbaum, M.A. Klingberg and E. Shenberg, Am. J.
    Epidemiol., 91(1970)52.

49 R.M. Minzat and V. Tomescu, Arch. Exp. Veterinaermed., 29(1975)557.

50 P.H. van Thiel, Geneesk Tijdschr. v Neder Indie, 78(1938)1859.

51 L. Clark, J.I. Kresse, R.R. Marshak and C.J. Hollister, J. Amer. Vet.
    Med. Assoc., 141(1962)710.

52 C.E.G. Smith and L.H. Turner, Bull. World Health Organ., 24(1961)35.

53 A.T. Laird, G.L. Kite and D.A. Stewart, J. Med. Res., 29(1913)31.

54 F.J.W. Miller and J.P. Anderson, Arch. Diseases Childhood, 29(1954)152.

55 L. Coin, M.L. Menetrier, J. Labonde and M.C. Hannoun, Second
    International Conference on Water Pollution Research, Pegamon, N.Y.,
    1964(1966).

56 C. Rhines, Am. Rev. Tuberculosis, 31(1935)493.

57 C.H. Bullin, E.J. Tanner and C.H. Collins, J. Hyg., 68(1970)97.

58 D.A. McSwiggan and C.H. Collins, Tubercle, 55(1974)291.

59 R.K. Bailey, S. Wyles, M. Dingley, F. Hesse and G.W. Kent, Am. Rev.
    Respir. Dis., 101(1970)430.

60 S. Goslee and E. Wolinsky, Am. Rev. Respir. Dis., 113(1976)287.

61 J.W. Moseley, in G. Berg (Editor), Transmission of Viruses by the Water
    Route, Wiley-Interscience, New York, 1967, p. 84.

62 G. Papaevangelou, A. Kyriakidou, G. Vissoulis and D. Trichopoulos,
    J. Hyg., 76(1976)229.

63 R. Viswanathan, Int. J. Med. Res., 45(1957)Suppl. 1.

64 J.A. Bell, W.P. Rowe, J.I. Engler, R.H. Parrott and R.J. Huebner,
    J. Am. Med. Assoc., 157(1955)1083.

65 H.M. Foy, M.K. Cooney and J.B. Hatlen, Arch. Environ. Health, 17(1968)
    795.

66   S.R. Weibel, F.R. Dixon, R.B. Weidner and L.J. McCabe, J. Am. Wat.
       Works Assoc., 56(1964)947.

67   H.O. Lobel, A.L. Bisno, M. Goldfield and J.E. Prier, Am. J. Epidemiol.,
       89(1969)384.

68   R. Dolin, N.R. Blacklaw, H. Dupont, S. Farmal, R.F. Buscho, J.A. Kasel,
       R. Chames, R. Hornick and R.M. Chanock, J. Infect. Dis., 123(1971)307.

69   J. Gomez-Barreto, E.L. Palmer, A.J. Nalmias and M.J. Hatch, J. Am. Med.
       Ass., 235(1976)1857.

70   J.A. Kasel, L. Rosen and H.E. Evans, Proc. Soc. Exp. Biol. Med., 112
       (1963)979.

71   D.L. Lundgren, W.E. Clapper and A. Sanchez, Proc. Soc. Exp. Biol. Med.,
       128(1968)463.

72   N.A. Clarke, R.E. Stevenson and P.W. Kabler, J. Amer. Water Works
       Assoc., 48(1956)677.

73   E. Lycke, S. Magnusson and E. Lund, Arch. Ges. Virusforsch, 17(1965)409.

74   M.L. Peterson, Proc. 10th Conf. Great Lakes Res., 16(1967)79.

75   G. Joyce and H.H. Weiser, J. Amer. Water Works Assoc., 59(1967)491.

76   R.B. Burrows, Amer. J. Trop. Med. Hyg., 10(1961)172.

77   C.A. Le Maistre, R. Sappenfield, C. Culbertson, F.R.N. Carter,
       A. Offutt, H. Black and M.M. Brooke, Am. J. Hyg., 64(1956)30.

78   H. Kott, N. Buras and Y. Kott, J. Protozoal Suppl., 13(1966)33.

79   W.C. Boeck, J. Hyg., 1(1921)527.

80   S.L. Chang, Am. J. Hyg., 61(1955)103.

81   G.T. Moore, W.M. Cross, D. McGuire, C.S. Mollohan, N.N. Gleason,
       G.R. Healy and L.H. Newton, New England J. Med., 281(1969)402.

82   R.E. Bradsky, H.C. Spencer and M.G. Schultz, J. Infect. Dis., 130(1974)
       319.

83   R.J. Duma, H.W. Ferrell, E.C. Nelson and M.M. Jones, New England J.
       Med., 281(1969)1315.

84   L. Cerva, K. Novak and C.G. Culbertson, Am. J. Epidemiol., 88(1968)436.

85   M. Fowler and R.F. Carter, Br. Med. J., (ii)(1965)740.

86   J. de Jonckheere and H. van de Voorde, Appl. Environ. Microbiol., 31
       (1976)294.

87   B.H. Dean and A.E. Greenberg, Sewage Ind. Wastes, 30(1958)262.

88   P.H. Silverman and R.B. Griffiths, Ann. Trop. Med. Parasit., 49(1955)
       436.

89   H. Sinnecker, Z. Ges. Hyg., 4(1958)98.

90   W. Baumhogger, Z. Hyg. Infekt. Kr., 129(1949)488.

91 W.L.L. Wang and S.G. Dunlop, Sewage Works J., 26(1954)1031.

92 P. Jordan, L. Woodstock, G.O. Unran and J.A. Cook, Bull. W.H.O., 52 (1975)9.

93 A.M. Sologub, Hyg. Sanitation, Moscow, 11(1957)13.

WATER TESTING

The numbers of micro-organisms found in water vary depending on the source. When rain falls it becomes contaminated by collecting dust particles, but the bacterial numbers can be as low as 10-20/l of rain. Snow and hail are less pure, probably due to their larger surface area collecting a larger number of dust particles. The level of bacteria in ice can be quite high, and their survival is good.

The upland rivers are usually relatively pure, and the bacterial count is mainly derived from the soil. The presence of humic acid, from decaying vegetation, can lower the pH and cause the death of bacteria. Further downstream the rivers can become heavily contaminated and will include large numbers of micro-organisms derived from sewage. If the water is discharged into a lake, there is a continuous process of self-purification as bacteria are adsorbed onto particles and deposited, or ingested by protozoa. Similar processes occur when water is impounded in reservoirs, and a process of equalization takes place if the water is taken from several sources.

Shallow wells can be heavily contaminated containing up to 20,000 organisms/ml, and regular testing must be undertaken to ensure that the water is microbiologically safe. Water taken from deep wells is probably the purest. The bacteria are removed during percolation through the first 15 ft of moderately dense soil, and the numbers are reduced even more if the water is filtered through thick strata.

Water Sampling

The two major difficulties encountered in assessing the quality of surface waters, sewage or sewage effluents, and industrial effluents are that the composition of most of these liquids may fluctuate rapidly over wide limits, and the composition of samples may change before they can be analysed. This is particularly true where the chemical composition is being examined. Although there is variation in the pattern of discharge of domestic and industrial effluents, with batch processes showing the largest variation, there is little variation in the composition of the final effluent.

After sampling, rapid changes can occur in the concentration of many of the constituents, and so the sampling and analysis have to be designed to take this into account. Where the concentration of dissolved gases, e.g. oxygen or carbon dioxide is the main interest, they must be determined as soon as possible after sampling as their concentration can be altered by exchange with the atmosphere.

When sewage is sampled there may be oxidation of ammonia after sampling. This depends on the availability of oxygen and the temperature. For example, if the sample is kept for 24 h at 20°C before incubation in the Biochemical Oxygen Demand (BOD) test, and if oxidation is not limited by the availability of oxygen, then the value obtained for ammonia will be ca. 80% of the true value. Under the same conditions, but with storage temperatures of 12°C and 4°C during the 24 h period, the corresponding values for the BOD would be about 85 and 95% respectively. Therefore the common practice is to keep sewage samples at 4°C until they can be analysed. There can be difficulties if samples are taken by an automatic sampler during the night.

The access of air can be reduced by keeping the sample in a full and tightly stoppered bottle. The bacterial activity will continue, consuming any oxygen present in solution, after which soluble nitrate may be reduced to nitrogen gas. Any organic material present may be digested anaero- bically to give methane and carbon dioxide. All these processes occur more rapidly at higher storage temperatures.

If only the COD (Chemical Oxygen Demand), or permanganate value, or the level of nitrogenous substances or surface active agents is required, then mercuric chloride can be added as a sterilizing agent. This can be added at 3 p.p.m. for good sewage effluent, but 50 p.p.m. may be required for samples with a high concentration of sewage. The addition of a sterilizing agent is of no value if the BOD of the sample is required. When unsteri- lized samples are to be stored, they should be kept in the dark prior to analysis, because any algae present would otherwise increase the concen- tration of dissolved oxygen by photosynthesis.

The samples may be taken from sewers or open channels and this can be done with simple apparatus e.g. polythene beakers. Here, the quality of the sample depends upon the suspended solids and so a true proportion of these must be included. There must be sufficient turbulence at the sampling position to keep the solids in suspension, and care must be taken to avoid scraping growths or deposits from the side. It is often difficult to obtain a true sample if oil is present.

Bacteriological Sampling

Every care must be taken to prevent contamination of the water sample. Specimens can be taken in bottles of ca. 250 ml capacity, with ground glass stoppers having an overhanging rim. The bottle should be sterilized by autoclaving and then dried at 150°C. Screw-capped bottles wrapped in kraft paper and sterilized by autoclaving may also be used. If the water supply which is being examined has been chlorinated, then 1-2 crystals of sodium thiosulphate should be added to the bottle prior to sterilization.

The process of opening and closing of the bottle to collect the sample must be carried out with particular care, to avoid any accidental bacterial contamination of the sample. If the sample is to be taken from a tap, then the mouth of the tap must be flamed and the water allowed to run for 5 min before the sample is collected in the bottle. To sample natural sources of water, the stopper is removed carefully, the bottle held at its base and inserted mouth downward a foot below the surface, the bottle is then turned so that the current flows into the opening without coming into contact with the hand. If the source is still e.g. a lake, then the bottle is moved horizontally, opening foremost. The bottle is brought to the surface and closed, the stopper must not be contaminated. The sampling of surface water should be avoided because this may contain decaying vegetation.

If the sample is required from a specific depth, then a weighted bottle should be used and unstoppered at the correct depth. Sediment deposits may provide a stable index of the general quality of the overlying water, particularly when there is a great variation in the bacterial quality of the water. A bottom sampler may be required for this and samples have been taken from depths of 25 m into sterile plastic bags [1].

The sample should be packed in ice and the tests carried out as soon as possible. A note must be made of the source of the sample, the geological nature of the ground and the proximity of any sources of contamination. A single laboratory examination is not sufficient, the water must be sampled frequently.

Before any tests are carried out the water sample must be thoroughly mixed. The bacteriological tests can be used to evaluate the quality of water from any source.

The concept of testing for faecal coliforms and faecal streptococci and their application in tropical waters have been questioned because they were designed and tested mainly in temperate climates. This problem was

examined in the streams of the Saka Valley, New Guinea [2]. Faecal coliforms and faecal streptococci were excellent indicators of the faecal
pollution which derived mainly from herds of domestic pigs. The water
temperatures ranged from 13-26.2°C but there was no evidence of coliform
regrowth, probably because no major discharges of domestic or industrial
waste were present.

Frequency of Sampling Water Supplies for Bacteriological Examination

The bacteriological quality of the water supply must be examined at
frequent intervals. In conjunction with this it is advisable to check the
chlorine level in the water. Chlorinated supplies will require bacteriological examination daily as it enters the distribution system. This can
be carried out for the larger supplies, but where a population of 10,000
or less is served it is not really practical. Here, reliance is usually
placed on the residual chlorine levels and the bacterial quality of the
water examined once a week.

When naturally pure waters are distributed without chlorination, the
frequency of examination is based on the population served and the
character of the source viz.:

Population Served	Maximum Interval Between Successive Samples
Less than 20,000	1 month
20,000-50,000	2 weeks
50,000-100,000	4 days
Over 100,000	1 day

The water should be sampled at all the points at which it enters the
distribution system.

Both chlorinated and unchlorinated supplies can undergo deterioration
in the distribution system, and so there is no difference between them in
the frequency of sampling. Where the population is under 100,000 persons
one sample/5,000 of the population served should be examined each month;
populations of 100,000-500,000, one sample/10,000/month; and for populations over 500,000, one sample/20,000/month. These samples should be
spaced out evenly over the month [3,4].

Small rural supplies should be sampled at monthly intervals, or less. If this is not practical, then it is not really possible to state that a supply is satisfactory. When a new well, or spring, is brought into use the water should be sampled at frequent intervals under a variety of climatic conditions to monitor any variation in the quality of the water.

Bacteriological Testing of Water

The use of colony counts gives an indication of the bacteriological quality of the water and is of value when the water is to be used in the preparation of food and drink. Colony counts can give an advance warning of pollution. However, the practice in the United Kingdom is to concentrate on testing for the presence of Escherichia coli or coliform organisms in the water.

Plate Count

A 1 ml sample of the water is taken with a sterile pipette and placed in a sterile 9 cm petri dish and 10 ml of molten, cooled yeast extract agar is added. The whole is mixed thoroughly and allowed to solidify. If the water is suspected of being contaminated it can be diluted 1:10 and 1:100 with sterile $\frac{1}{4}$ strength Ringer's solution. Then 1 ml aliquots of these dilutions are taken and mixed with agar as before. Duplicate plates are prepared from each dilution and one plate is incubated at 37°C for 24-48 h and the other at 20-22°C for 66-72 hr. After incubation the colonies which have developed in the medium are counted, a hand lens may help in the counting of small colonies. The results are reported as the number of colonies developing from 1 ml of the original sample, because each colony does not necessarily come from one organism. Only plates with between 30 to 300 colonies are counted. Values below 30 are not statistically significant and values over 300 colonies/plate are not accurate due to overcrowding and competition.

A high count at 37°C after 24 hr incubation indicates faecal pollution; if the plate is left for 48 h saprophytes may grow and give a high result. The ratio of the counts at 22°:37° is generally 10 or more for unpolluted waters and pollution is indicated if this ratio is less than 10. For example, water taken from a deep well may give a count of 5-10/ml at 37°C and 1,000/ml at 22°C. The organisms in the latter case are mainly chromogenic bacteria and the water is bacteriologically satisfactory.

Examination for Coli-aerogenes Group of Organisms

The science of sanitary water bacteriology began in 1880 when von Fritsch described Klebsiella pneumonia and K. rhinoscleromatis as organisms which were characteristic of human faecal contamination. A short time later, Escherich identified Bacillus coli as an indicator of faecal pollution. Both workers considered that human faeces were a dangerous source of pollution whilst the faeces of other warm-blooded animals were not considered a health hazard.

Water which is required for human consumption must be free from toxic chemicals and from pollution by human or animal excreta and therefore free from pathogenic micro-organisms. In most communities this will necessitate some form of artificial purification. Such processes have to be checked by a routine programme of examination of the final product before distribution. The rationale and the techniques used in the bacteriological examination are discussed in the Ministry of Health Report No. 71 [5].

The detection of faecal pollution is of particular importance because water borne infections are still an important source of disease. The protection of water supplies is a continuing obligation of the public health authorities. Only a few pathogens reaching water from faecal contamination will survive for more than a short period.

Escherichia coli is found regularly in the intestines of all animals, including man, and when large numbers of this organism are found in water recent pollution is suggested. Tests for this organism are carried out as it can be detected at 1/100 ml of water and is thus a very sensitive indicator. Low numbers of E. coli indicate a more remote contamination since it is assumed to have a short half-life when separated from the host. However, the evidence for the survival of E. coli in water is conflicting [6].

The presence of E. coli, which is a Gram-negative non-spore forming rod producing acid and gas by fermentation of lactose, is the cardinal index of faecal pollution. Certain other organisms with these characteristics are also found in water, but they are non-faecal in origin. The whole group is referred to as the coli-aerogenes group (coliform organisms). The examination of a water sample is based firstly on the detection of the coli-aerogenes group in a presumptive coliform test, and secondly on the confirmation of the presence of E. coli in the differential coliform test.

There are different views on whether a presumptive test for the presence of coliform organisms should detect E. coli biotype I (which produces acid and gas from lactose, and indole production at 44°C), or it should be based solely on acid and gas formation from lactose at 44.0 to 44.5°C. The latter criteria are the basis of the North American test which will detect E. coli type I and irregular types II and VI [7] which may not be of faecal origin [8]. The European approach has been to base recognition of presumptive faecal coliform organisms on acid plus gas and indole formation at 44°C [8] which excludes the irregular types and will detect only E. coli biotype I.

These tests also indirectly examine for the presence of Salmonella typhi and other water borne pathogenic bacteria, but do not give a reliable indication of the survival or presence of potentially pathogenic viruses.

The Presumptive Test for the Coli-aerogenes Group

An estimation of the number of coliform organisms in a water supply can be made by adding varying quantities of the water, from 0.1-50 ml to MacConkey's liquid medium. The single strength medium has the composition:-

MacConkey Bile-salt Lactose Peptone Water (g/l)

Sodium taurocholate (commercial)	5 g
Peptone	20 g
Sodium chloride	5 g
Lactose	10 g
Bromocresol purple, 1% solution in ethanol OR	1 ml
Neutral red	5 ml
Distilled water to	1 litre

The single strength medium is distributed in 10 ml amounts into bottles containing Durham tubes and sterilized. Also double strength medium is prepared i.e. the quantity of all ingredients, except the water is doubled, and distributed in 50 ml and 10 ml amounts into bottles containing Durham tubes. The Durham tubes must be full of liquid so that the production of gas can be detected. The size of the bottle used varies with the quantity of medium and the volume of water which has to be added to it.

For good quality waters it is sufficient to examine aseptically (using sterile pipettes) one 50 ml sample, five 10 ml samples and five 1 ml samples and to add these to a 50 ml bottle, and five 10 ml bottles of double strength MacConkey broth, and to five 10 ml bottles of single strength medium respectively. For water taken from sources thought to be polluted, a further five 0.1 ml samples of water may be added to five

10 ml bottles of single strength MacConkey broth. The inoculated bottles are then incubated at $37^{\circ}C$ and examined after 18-24 h.

The test relies on the fact that the coli-aerogenes group of organisms will ferment lactose to give acid and gas. The production of acid is shown by a change in colour of the indicator to yellow and gas is seen in the Durham tube. There must be sufficient gas to fill the concavity at the end of the Durham tube before a positive result can be recorded. The presence of bile salts suppresses growth of Gram-positive organisms. Any samples which are negative are reincubated for a further 24 h. Bile salts and peptone can vary in quality and Teepol 610 has been suggested as an alternative to bile salts as an inhibitor [9].

Another alternative to MacConkey broth in the presumptive test is an improved formate-lactose-glutamate medium [10]. The production of acid and gas in this medium when inoculated with unchlorinated water and incubated at $37^{\circ}C$ for 48 h indicates the presence of coliform organisms. However, some spore-forming aerobic organisms can cause false presumptive reactions. The medium below is regarded as being very satisfactory for general use.

Improved Formate Lactose Glutamate Medium (Double Strength)

Lactose	20.0	g
L(+) glutamic acid (sodium salt)	12.7	g
L(+) arginine monohydrochloride	0.04	g
L(-) aspartic acid	0.048	g
L(-) cystine	0.04	g
Sodium formate	0.5	g
K_2HPO_4	1.8	g
Ammonium chloride	5.0	g
Magnesium sulphate ($7H_2O$)	0.2	g
Calcium chloride ($2H_2O$)	0.02	g
Ferric citrate scales	0.02	g
Thiamin	0.002	g
Nicotinic acid	0.002	g
Pantothenic acid	0.002	g
Bromocresol purple	0.2	g
Distilled water to	1 litre	

Kampelmacher et al. [11] found that formate-lactose-glutamate medium as an enrichment technique, was the best method for the enumeration of the coli-aerogenes group in water. The use of Eijkman lactose broth at $44^{\circ}C$ was better for E. coli than Eijkman glucose broth, but for an examination of the combination of E. coli and the coli-aerogenes group the formate-lactose-glutamate medium was the best of all. They also reported that they obtained the same results if the water samples were stored at $4\text{-}6^{\circ}C$ for 24 h or tested immediately.

In reporting the results of the presumptive test reference is made to McCrady's probability tables [5]. The various combinations of positive and negative results allow the probable number of coliform organisms in 100 ml of the water sample to be estimated.

The Differential Coli-aerogenes Test

The formation of acid from lactose in media containing bile salts after incubation at $44^{\circ}C$ can be used for direct plate count or detection of E. coli [12,13]. To determine whether the coliform organisms which were detected in the presumptive test are E. coli then further differential tests must be applied. The one test which invariably is applied is the Eijkman test. This depends on the ability of E. coli biotype I to produce gas when grown in MacConkey medium at $44^{\circ}C$. All the tubes which show positive acid and gas production in the presumptive test are inoculated into fresh bottles of single strength MacConkey's medium; it is preferable to incubate these bottles at $37^{\circ}C$ before inoculation. The tubes are then incubated for 24 h at $44^{\circ} \pm 0.5^{\circ}C$ in a water-bath, an incubator is not satisfactory for this purpose. Those samples which show production of gas may be regarded as containing E. coli and an estimation of the number in 100 ml can be made as before. The saprophytes which may have been in the water sample and given acid and gas at $37^{\circ}C$ cannot grow at $44^{\circ}C$, which is near the upper limit for E. coli. An alternative medium for use in the Eijkman test is brilliant green bile broth in bottles which contain a Durham tube.

Brilliant Green Bile Broth

Peptone	10 g
Ox bile (purified)	20 g
Lactose	10 g
Brilliant green, 0.1% aqueous solution	13 ml
Distilled water to	1 litre

There seems to be no difference between the results obtained with any of the commercially pre-prepared brilliant green bile broths [14].

The presence of brilliant green suppresses the growth of anaerobic lactose fermenters e.g. Clostridium perfringens which could give false positive results at $44^{\circ}C$. It has been shown that a 1% lactose ricinoleate broth gave better results than brilliant green broth and was subject to less variability than the bile or brilliant green [15].

The other tests used for the differentiation of the coli-aerogenes group are those making up the IMViC series. The first is based on the ability of E. coli to form free indole from tryptophan:-

The tubes showing positive results in the presumptive coliform tests are subcultured into a peptone broth.

Peptone Broth

Peptone, which contains sufficient tryptophan	20 g
Sodium chloride	5 g
Distilled water to	1 litre

After incubation for 48 h at 37° a test is applied for free indole by adding 0.5 ml of Kovac's reagent which is layered on the top. (Kovac's reagent consists of amyl alcohol, 150 ml, p-dimethylaminobenzaldehyde, 10 g, concentrated hydrochloric acid 50 ml.) A red colour in the alcohol layer indicates a positive reaction. If incubation of this test is carried out at $44^{\circ}C$, typical coliforms can form indole whilst other 'irregular' types cannot.

The use of lactose fermentation alone to detect E. coli will miss

those strains which are unable to ferment lactose [16] to produce acid in
two days. Enteropathogenic E. coli may occasionally be late or non-lactose
fermenters [17]. Only 90% of Escherichia strains will produce acid from
lactose within two days, whilst 99% of all strains will produce indole
[18]. The ability to produce indole at 44° has been used in the development of direct plate counts and detection of E. coli in water or food [19].
Lactose tryptone ricinoleate medium can be used at 44°C for E. coli
detection in 24 h [20].

The second test which is applied requires subculture of the positive
presumptive tests into glucose-phosphate broth:-

Glucose-phosphate-peptone Water

Peptone	5 g
Dipotassium hydrogen phosphate	5 g
Distilled water to	1 litre
Glucose, 10% solution, sterilized separately	50 ml

The amount of acid produced when coliforms ferment glucose varies and this
broth is highly buffered with dipotassium hydrogen phosphate at pH 7.5. If
sufficient acid is produced the buffering capacity is overcome and the pH
falls to less than 4.6. This can be detected by adding about five drops of
methyl red indicator solution, after the bottles have been incubated for
48 h at 37°. Positive results, E. coli, are bright red and negative
results are yellow. If the results obtained after 48 h are equivocal, the
test should be repeated with cultures that have been incubated for five
days.

The glucose-phosphate broth culture may also be used for the Voges-
Proskauer test which depends on the fact that many bacteria ferment carbohydrates with the production of acetyl methyl carbinol (acetoin,
$CH_3CO.CHOH.CH_3$) or its reduction product 2;3-butylene glycol (2;3-butane-
diol, $CH_3CHOH.CHOH.CH_3$). These substances can be tested by using a colori-
metric reaction between diacetyl ($CH_3CO.CO.CH_3$), and a guanidino group,
under alkaline conditions. Diacetyl is formed during the test by oxidation
of acetyl methyl carbinol or 2;3-butylene glycol.

The test may be carried out on a glucose-phosphate broth culture after
48 h incubation at 37°C. To this culture is added 1 ml of a 40% solution
of potassium hydroxide and 3 ml of a 5% solution of α-naphthol in ethanol.

A positive reaction is indicated by the development of a pink colour in 2-5 min which becomes crimson in 30 min. To ensure maximum aeration the tubes are shaken from time to time. Alternatively O'Meara reagent may be used (40 g potassium hydroxide and 0.3 g creatine in 100 ml distilled water) but an incubation time of up to 4 h at $37^{\circ}C$ may be required for the production of an eosin pink colour, although this may appear in 2-5 min. This reagent removes the need to use α-naphthol which is carcinogenic. An organism of the enterobacterial group is usually either methyl-red positive and Voges-Proskauer negative or methyl-red negative and Voges-Proskauer positive.

The final test is to examine the organism's ability to utilize citrate as a sole source of carbon. The medium employed may be either Koser's liquid citrate medium or Simmon's citrate agar.

Koser's Medium

Sodium chloride	5.0 g
Magnesium sulphate	0.2 g
Ammonium dihydrogen phosphate	1.0 g
Potassium dihydrogen phosphate	1.0 g
Sodium citrate	5.0 g
Distilled water to	1 litre
pH 6.8	

Simmon's Medium

Koser's medium	1 litre
Agar	20 g
Bromothymol blue 0.2%	40 ml

This is dispensed into 10 ml amounts, autoclaved at $121^{\circ}C$ for 15 min and allowed to set as slopes.

The error commonly encountered in the use of this medium is the employment of too large an inoculum. It is better if the organism is introduced into water and this used to inoculate the tubes. Citrate can be used by K. aerogenes as a sole source of carbon but cannot be utilized as sole carbon source by E. coli of faecal origin. Only 7% of organisms

isolated from faeces will give a positive reaction whereas 90% of coli-
aerogenes organisms from soil are citrate positive. The results obtained
on Koser's citrate medium can be interpreted as follows: growth (turbidity)
is positive; no growth, negative. Results obtained on Simmon's citrate
medium are as follows: blue colouration of medium and growth, positive;
original green colour and no growth, negative.

The results obtained may be summarized:-

	Indole	Methyl red	Voges-Proskauer	Citrate
E. coli	+	+	-	-
K. aerogenes	-	-	+	+

The additional tests may give a further aid:-

	Indole Production at 44°C	Gas in Brilliant Green Bile Broth at 44°C	Eijkman Test at 44°C
Typical coliform organisms	+	+	+
'Irregular' types of coliform organisms	-	+	+
Other organisms)	+	-	-
)	-	-	-

There are many organisms which give a variety of results between those
obtained with E. coli of faecal origin and the saprophytic K. aerogenes.
These are wrongly termed the Intermediate coliforms. They are not intermediate between K. aerogenes and E. coli but are similar to E. coli,
although they are not of faecal origin and more tests would be required to
classify them. For example, two types of atypical coliform organisms will
give gas at 44°C in the Eijkman test but are unable to produce indole at
44°C. They occur only rarely in water supplies in the U.K., although the
joints of water mains may be packed with imported jute which may be contaminated with them. These were classified as Irregular Type II and IV
and these terms are still retained in water bacteriology because complete
identification is unnecessary.

Further bacteriological tests can be carried out for the presence of
Clostridium perfringens (welchii) because the spores of this organism may
survive chlorination. Clostridium perfringens ranges from 56-71% of the
total Clostridium populations of wastewater samples, but represents only

0.4-4.1% of the total group in freshwater sediments and soil samples [21].

In this test 50 ml of the water sample is added to 100 ml of sterile milk in a stoppered bottle. This is then heated at 80°C for 15 min, to destroy the non-sporing organisms, and sterile liquid paraffin is layered on the surface to maintain anaerobic conditions. The bottles are incubated at 37°C for five days. The 'stormy clot' reaction which is indicative of the presence of Cl. perfringens (welchii) can develop within 24-72 h. The probable number of Cl. perfringens can be estimated using varying quantities of water, as in the presumptive coliform test. Differential reinforced clostridial medium may be used to isolate Cl. perfringens from water more frequently than the 'stormy clot' reaction. For this test 50 ml of water is added to 50 ml of double strength differential reinforced clostridial media and then incubated at 37°C for 48 h.

Differential Reinforced Clostridial Medium (D.R.C.M.)

Peptone	10.0 g
Lab Lemco	10.0 g
Sodium acetate (hydrated)	5.0 g
Yeast extract	1.5 g
Soluble starch	1.0 g
Glucose	1.0 g
L(-) cysteine	0.5 g
Distilled water to	1 litre

Prior to use add, aseptically, 0.04% w/v anhydrous sodium sulphite and 0.07% w/v ferric citrate.

A positive reaction is shown by a blackening of the medium and one loopful from such positive bottles is inoculated into sterile milk, and then incubated at 37°C for 48 h. The production of a stormy clot confirms the presence of Cl. perfringens (welchii). The long period of survival of the spores of this organism, which occurs in much smaller numbers than E. coli in recently contaminated water, is of value in detecting earlier pollution of the water. In addition, it helps in the confirmation of the faecal origin of atypical coliforms in the absence of E. coli. Long standing contamination of shallow and surface well water may give a positive test for Cl. perfringens, even if coliforms are absent.

Tests for Faecal Streptococci

The term 'enterococci' which is often used synonymously with faecal streptococci is not precisely defined and so should be avoided. The faecal streptococci are those which normally occur in human and animal faeces. The species concerned belong to Lancefield's serological group D and include <u>Streptococcus faecalis</u>, <u>S. faecium</u>, <u>S. durans</u>, <u>S. bovis</u> and <u>S. equinus</u>. The properties of faecal streptococci which are used in their detection are: the ability to grow at 45°C, to grow in the presence of 40% bile salts and to grow in the presence of concentrations of sodium azide which are inhibitory to coliforms and other Gram-negative organisms.

Preliminary resuscitation at 37°C is often required before growth at 45°C, since the ability to grow at this temperature may be temporarily lost once the organism is outside the body. The numbers of faecal streptococci may be estimated by a multiple tube method similar to that used for the coli-aerogenes. The required volumes of water are inoculated into tubes of single or double strength glucose azide broth.

Glucose Azide Broth

Peptone	10.0	g
Sodium chloride	5.0	g
K_2HPO_4	5.0	g
KH_2PO_4	2.0	g
Glucose	5.0	g
Yeast extract	3.0	g
Sodium azide	0.25	g
Bromocresol purple	0.032	g
Distilled water to	1 litre	

Durham tubes are not required in this test because there is no production of gas. The tubes are then incubated at 37°C for 72 h. Once acid production is observed a heavy inoculum is subcultured into fresh tubes of glucose azide broth. These are then incubated at 45°C for 78 h. The tubes which show production of acid at this temperature contain faecal streptococci.

Membrane Filtration Technique for the Examination of Water and Sewage

This technique is based on the use of a porous cellulose acetate membrane. The pore structure of the membrane allows large volumes of water to be filtered rapidly, under pressure, and prevents the passage of bacteria. The bacteria are sieved out and remain on the surface of the membrane. When the membrane is placed on a nutrient medium, the nutrients can diffuse through the pores and the organisms present will grow and produce surface colonies. The cellulose acetate pore structure may limit the diffusion of nutrients and low counts can result [22].

The amount of water tested depends on the likely degree of pollution, should there be any doubt then two dilutions should be examined. For purified tap water 250-500 ml should be tested, for well water 10 and 100 ml and for river water 1 and 10 ml. Any volume less than 20 ml should be made up to that volume before being passed through the filter.

One advantage of membrane filtration is that it can concentrate very dilute suspensions of organisms and by the use of selective media very low levels of pathogenic organisms can be enumerated. Thus, Wilson and Blair's bismuth sulphite medium can be used to detect typhoid and paratyphoid bacilli.

Misleading results can be obtained with membrane filters, particularly when tests for faecal organisms are carried out at elevated temperatures. When autoclaved membranes were examined for toxicity, they were capable of recovering 92% of the E. coli at 35°C, but no more than 40% when the incubation temperature was 44.5°C [23]. The membranes may have hydrophobic areas which limit the effective filtering area or the pH may be lowered by ethylene oxide sterilization.

After filtration the membranes are placed on either a nutrient agar or on sterile pads soaked in a nutrient broth. The nutrients diffuse through the membrane and the colonies develop on the surface. Counting is carried out either by visual inspection or under a low-power binocular microscope.

A nutrient broth of good quality is required to give counts which are comparable with pour plate counts e.g.

M-yeast Extract Broth

Yeast extract	6 g
Peptone	40 g
Distilled water to	1 litre
pH 7.4	

A modification of the McConkey broth can be used to estimate the number of coliform organisms and the count can be made in 18 h. There is no need for a resuscitation medium.

Teepol Lactose Medium

Peptone	20 g
Lactose	10 g
Sodium chloride	5 g
Teepol 610	1 g
Bromothymol blue (0.2% w/v solution)	25 ml
Distilled water to	1 litre

The Teepol (a detergent containing sodium and potassium salts of alkyl sulphates) is used in place of the bile salts in McConkey medium. It is claimed to be cheaper and more reliable. The lactose-fermenting E. coli colonies are pale cream in colour, large and opaque.

The numbers of E. coli biotype I in foods can be estimated within 24 h if the production of indole at 44° is used. The membrane is placed on tryptone bile agar after filtering the suspension or organisms [24].

Membrane filtration can be used to detect faecal streptococci in water samples. After filtration the membrane can be placed on a well dried plate of Slanetz and Bartley's medium.

Slanetz and Bartley's Glucose Azide Agar

Tryptose	20.0 g
Yeast extract	5.0 g
Glucose	2.0 g
K_2HPO_4	4.0 g
Sodium azide	0.4 g
Agar	12.0 g
Distilled water to	1 litre

After sterilization, add 10 ml of a 1% w/v solution of 2,3,5-triphenyl-tetrazolium chloride (T.T.C.) and pour plates.

The medium is incubated at $37^{\circ}C$ for 4 h and then at $45^{\circ}C$ for 44 h. All red or maroon colonies are counted as faecal streptococci. These colonies may be confirmed as faecal streptococci by direct microscopical examination or by sub-culture on MacConkey agar, where they appear as small red colonies after 24-48 h incubation at 37°C.

An alternative medium for confirmation is a tyrosine sorbitol thallous acetate agar. Cultures are plated onto this medium and incubated at 45°C for three days. This medium can be used for direct plating of samples or for membrane filtration if an imprinting technique is used.

Tyrosine Sorbitol Thallous Acetate Agar

Peptone	10 g
Yeast extract	1 g
Sorbitol	2 g
Tyrosine	5 g
Agar	12 g
Distilled water to	1 litre

Dissolve and sterilize by autoclaving, then add 10 ml of a 1% w/v solution of 2,3,5-triphenyltetrazolium chloride (T.T.C.) and 1 g of thallous acetate. Pour this as a thin layer in a petri dish. When set pour another thin layer of medium to which an extra 4 g/l of tyrosine is added to give a suspension.

Streptococcus faecalis produces uniformly dark maroon colonies surrounded by a clear zone. The medium takes advantage of the fact that S. faecalis will ferment sorbitol to produce acid and it also produces a tyrosine decarboxylase which is only effective in acid conditions (produced by the fermentation of sorbitol). The decarboxylation of tyrosine gives the clear zones around the colonies and the maroon colour comes from the reduction of T.T.C. at pH 6.2. Pfizer selective enterococcus medium is highly selective for enterococci when membrane filtration is used [25].

A comparison between the most probable number methods and the membrane filtration technique indicated there was no significant difference in the accuracy of the techniques, but the membrane filter method was more precise [26]. When the methods were examined for detecting coliforms and faecal streptococci in chlorinated sewage effluent, the multiple tube technique gave better results than the direct membrane filter technique [27]. The preliminary enrichment (the presumptive test) before transfer to a selective medium (the differential test) may be better suited to the survival of stressed organisms.

Pathogenic Organisms in Water

The most important of the waterborne pathogens are the salmonellae and the most significant is Salmonella typhi, and to a lesser extent S. paratyphi B. These organisms can be recovered from river waters even in the order of 1 to 2/1. The techniques in current use are modifications of the method of Hammerström and Ljutov [28] and Ljutov [29]. A layer of diatomaceous earth, on a wire mesh support, is substituted for the membrane at the bottom of the filter. A membrane filtration apparatus with a wide funnel is preferred. A small amount of sterile distilled water is poured into the funnel and to this is added a quantity of 1% diatomaceous earth until the required thickness is reached. The pressure is turned on in the unit, and the measured volume of sample is added before all the sterile water has gone through. When the fluid has been sucked through, the paste can be tipped into 100 ml of Selenite F medium. This is incubated at $42^{\circ}C$ and after 18 h subcultures can be made on selective media.

Selenite F Broth

Sodium acid selenite	4.0 g
Peptone	5.0 g
Lactose	4.0 g
Disodium hydrogen phosphate	9.5 g
Sodium dihydrogen phosphate	0.5 g
Sterile water	1 litre

The selective media used for plating out the enrichment culture are usually Wilson and Blair agar, Salmonella Shigella (SS) agar, brilliant green agar, or deoxycholate agar.

Non-pathogenic Nuisance Organisms in Water

Earthy, mouldy or musty tastes can be produced in water by actinomycetes or the microfungi. The actinomycetes have affected whole supplies through growth in reservoirs or raw waters. The microfungi can grow inside pipes where the water is used only infrequently and the pipes have been warmed locally. The most convenient way to examine a water supply for the presence of microfungi is to use membrane filtration because large volumes of water can be sampled. The membrane can be placed on a suitable medium, and the colonies produced after incubation for seven days at $22^{\circ}C$ can be counted.

Martin's rose bengal streptomycin agar is a useful medium for the enumeration of fungi.

Martin's Rose Bengal Streptomycin Agar

Peptone	5.0 g
Glucose	10.0 g
Potassium dihydrogen phosphate	1.0 g
Magnesium sulphate	0.5 g
Rose bengal	0.35 g
Agar	10.0 g
Distilled water to	1 litre

Streptomycin is added, prior to use, to give a final concentration of 100 μg/ml.

The fungi associated with mouldy tastes take six to seven days to produce colonies and they may be present at levels of 10-100/100 ml of water.

The presence of actinomycetes can be shown by using a medium with chitin as the sole source of carbon and nitrogen in a simple salts solution. As the chitin is insoluble it cannot diffuse through a membrane so the filter membrane is placed face down on the agar medium, incubated for 4 h at 28-30°C and the culture re-incubated for five to ten days at 28°C or up to three weeks at 22°C.

Bacteriological Standards

Water which has been chlorinated should be free from coliform organisms, so it should not be possible to show the presence of coliform organisms in any 100 ml sample. If any samples are positive, after carrying out the confirmatory tests, then the process of purification and sampling must be examined in detail.

Whilst piped supplies of unchlorinated water are not recommended, small supplies still exist. The water which enters such a system should not contain demonstrable E. coli in 100 ml. If E. coli is absent, the presence of not more than 3 coliform bacteria can be tolerated in occasional samples from established supplies. Steps should be taken to remove the source of pollution if further samples taken at the same point show the presence of coliform organisms. When the coliform organisms persist, or

increase to numbers greater than 3/100 ml then the supply must be chlorinated.

Small rural supplies should be of the same bacteriological quality as
a piped supply, and the access of pollution to the supply should be
prevented. The coliform count from a shallow well (by attention to removal
of all sources of pollution and provision of good brick lining, coping and
cover) should be less than 1/10 ml. If this is not attained and E. coli is
frequently detected, then the source of supply must be condemned.

When samples are taken from the distribution system they should be
free from coliform organisms. In practice this cannot always be attained,
and the standard applied should ensure that throughout the year 95% of the
samples should not contain any coliform organisms or E. coli in 100 ml.
No single sample should contain more than 10 coliform organisms/100 ml, nor
should a sample contain more than 2 E. coli/100 ml. There should not be
1 or 2 E. coli/100 ml together with a total coliform count of 3 or more/
100 ml. When consecutive samples are taken, coliform organisms should not
be detectable in any two. If coliform organisms are found the supply
should be resampled. Where levels of 1 or 2 E. coli or 1-10 coliform
organisms/100 ml persist, then the source of pollution and its access to
the supply must be located and remedied.

Water which is used in canning factories for cooling purposes should
not contain more than 100 organisms/ml and preferably the water should be
chlorinated.

Bacteriological Control of Swimming Baths

The water in public swimming baths could become infected with pathogenic organisms which come from contaminated water entering the pool or
from the bathers. These organisms could give rise to gastro-enteritis,
otitis media, skin infections etc. unless the water is adequately treated.
Most swimming baths have a system which takes the water from the deep end,
filters, clarifies and chlorinates the water before returning it at the
shallow end. The amount of chlorine introduced into the water is controlled to give a level of 0.2-0.5 p.p.m.

The bacteriological examination of the water is the same as that for
chlorinated supplies of drinking water. The standards applied are those of
a high purity drinking water.

The oral and nasal bacteria collect in the surface film of fatty
substances which come from the skin and hair of the bathers. These

organisms are protected from the action of the chlorine and they could be responsible for the spread of viral and bacterial disease. Swimming baths should be provided with an overflow gutter into which the surface film can drain.

Rapid Techniques for the Detection of Bacteria

There is a need in many areas of bacteriology for a method which will detect bacteria rapidly with a high sensitivity and a low background of interference. These objectives have not yet been achieved and the available techniques have been summarised [30]. The cost per test of these methods is probably higher than using the conventional tests, but this may be acceptable in certain circumstances.

Fewer than 100 colony-forming units of several bacterial species of bacteria could be detected in 2 h or less by using ^{14}C-labelled substrates [31]. The organism grows to the micro-colony stage on membrane filters and the radio-labelled CO_2 is trapped with barium hydroxide and the radio-activity measured. A similar technique has been used which employs radio-labelled lactose and the CO_2 evolved is trapped by hyamine hydroxide and then counted [32]. The method was claimed to be able to detect the activity arising from 1-10 coliform organisms, at the micro-colony stage, within 6 h of incubation. A simpler device using the same principles but requiring only conventional counting devices and normal laboratory glassware has been suggested [33].

The use of radio-active phosphorus to measure microbial activity after 1 h of growth is possible [34]. This requires that the portion of meta-bolized phosphorus insoluble in trichloroacetic acid is measured.

When these techniques are applied to estimating the numbers of coliform organisms and faecal streptococci in a water sample, then the radio-labelled substrate should be incorporated into a selective medium. If this is not carried out then the substrate could be metabolized non-specifically by the total microflora.

When a micro-organism produces gas as an end-product of metabolism then it is possible to detect the molecular hydrogen which is produced [35]. The test can be carried out and for 10 organisms/ml gives a result in 7 h. If the cell density is greater a result can be obtained more rapidly.

A gas liquid chromatographic method as a presumptive coliform test has been suggested which is based on the presence of metabolically produced ethanol [36]. This gives the best results when there are initially

5 coliforms/ml or higher. At this level the minimum detection time was
9 h.

A combination of membrane filtration and immunofluorescence permits
the enumeration of faecal streptococci in 10-12 h [37]. Small numbers of
bacteria can be detected and identified in an hour using an indirect radio-
labelled antibody staining technique [38].

Viruses

The virus-in-water question is an important aspect of rational water
quality control. The potential hazards arising from the discharge of
viruses in sewage effluent and the subsequent contamination of watercourses
mean that the presence or absence of coliforms may no longer be an adequate
control. The suggested standard is that there should be no more than one
detectable infectious virus unit per 380 litres of water [39].

Viruses can only multiply within living cells and so living organisms
such as animals, chick embryos or tissue culture must be used for laboratory isolation. When enteric viruses are considered one of two types of
tissue culture may be used: primary tissue culture or continuous cell
culture. Primary cell cultures are obtained from tissues removed from the
animal, minced or de-aggregated by exposure to enzymes, e.g. the proteolytic
enzyme trypsin (0.25%). When the separated cells are placed inside a
suitable glass or plastic flask, tube or plate, together with a complex
nutrient medium, they will attach to the wall of the vessel and multiply
to form a monolayer of cells. These confluent cell sheets can be removed
from the culture flask by brief exposure to chelating agents or trypsin,
diluted and inoculated into additional culture vessels with fresh medium.
The new progeny which develop can no longer be termed primary explant cell
cultures, and are called secondary cell cultures.

Continuous cultures are produced in a similar manner, but instead of
a tissue from an organ being used, a tissue culture is used as the primary
source of the cells.

When a virus is inoculated into a tissue culture some of the cells
become infected. The virus will multiply within these cells and spread to
neighbouring cells. At this stage the infected cell usually changes
biochemically and morphologically and dies. The result is a slow process
of destruction of cells in the culture which is termed a cytopathic effect
(CPE). The spread of virus from cell to cell can be slowed down by adding
a layer of agar, which contains the tissue culture medium, over the cells.

As a result the CPE is limited to a small area, rather than being confluent, and looks like a small hole in the monolayer of cells. These holes are termed plaques. Usually a single plaque arises from a single infected cell resulting from one virus infectious unit. When the dilutions are correct this method can be used to estimate the number of virus infectious units in a suspension, in the same way as bacterial counts. The term plaque forming unit (PFU) was given to the lowest concentration of viruses which forms one plaque on a cell monolayer.

The CPE can be used to identify different viruses by the plaque size and shape. However, this is not a reliable method of identification because viruses belonging to different groups can give identical CPE. The only reliable final identification is by the use of specific antisera.

The quantitative determination of enteric viruses may be carried out by either of two methods. The first is the plaque assay method, where 0.3-1.0 ml of virus dilutions are inoculated into tissue culture plates or bottles, and the cells are then covered with a layer of agar. The plates are incubated at $37^{\circ}C$ in an atmosphere containing 5% CO_2. After incubation the plates are examined for the presence of plaques, which are counted and the number obtained for each dilution is recorded. The number of PFU in the original virus suspension is calculated and the virus concentration reported as PFU per ml or other unit of volume.

The second method is the tube assay method where groups of tissue culture tubes are inoculated with each dilution. Following incubation at $37^{\circ}C$ the tubes are examined for CPE. The lowest dilution of virus is then found which caused CPE in 50% of the tubes. The figure so obtained is termed the tissue culture infectious dose - 50% ($TCID_{50}$) value of the suspension. The virus concentration can be calculated as a most probable number (MPN) using the same data. The selection of the method used usually depends on the resources and past experience of the laboratory involved, but both techniques give accurate and reproducible results.

The problem encountered when isolating viruses from water is the low concentration present. The number may be 1 PFU per litre of water or less. If 0.1 ml of material is required to inoculate a tube, 0.3 ml for a plate and 1.0 ml for a bottle, then it requires the inoculation of at least 1,000 bottles or 10,000 tubes to detect one virus in a litre of water. Therefore, techniques for virus concentration have been developed.

One of the first methods used for the detection of viruses in water was the gauze pad method. In this gauze pads or pads filled with cotton or plastic foam sponges are placed in the water and left for at least one and up to seven days. The viruses are adsorbed or entrapped in the pad and this leads to a concentration of virus particles from flowing water. The pads can be treated with dilute sodium hydroxide solution to increase the pH to 8.0, which improves the elution of trapped viruses from the pads [40]. The liquid which is expressed from the pads can be tested for viruses. The method is not quantitative but has proved useful for detecting the presence of viruses in water and sewage [40].

The conventional culture medium can be prepared in a concentrated form to permit the addition of 10-60 ml of the water sample under examination [41]. Alternatively, large volumes of a maintenance medium can be prepared from the water under test and this medium can be inoculated with a cell culture [42]. This does not concentrate the virus suspension but allows from 10 to 20 times the normal volume of water to be tested.

Ultracentrifugation can be used to concentrate small suspended solids from liquids, but viruses require 60,000 g for 1 h for sedimentation. This technique has been used [43] but is probably too costly and complicated for general application.

Viruses can be concentrated by electrophoresis [44] as they are negatively charged at neutral pH values and so move towards the anode when a potential is applied.

When a virus suspension is placed in a cellulose dialysis bag which is immersed in polyethylene glycol (a hydrophilic agent) water is absorbed from the bag and viruses remain inside [45,46]. The suspension may be concentrated a hundredfold.

The phase-separation method for the concentration of viruses depends on the fact that when two polymers e.g. dextran sulphate and polyethylene glycol, are mixed a two-phase system is formed. When particles or macromolecules are introduced the particles are partitioned between the phases, dependent on the size and surface properties of the particles [47]. The technique can be used for virus concentration because they have an almost one sided distribution and a 100 to 200-fold increase can be obtained [48], and enteroviruses were found in contaminated well water during an epidemic [49]. The dextran sulphate can be removed after the separation by the addition of barium or potassium ions.

Viruses can also be absorbed from water onto a wide variety of particulate material e.g. calcium phosphate [50], insoluble polyelectrolytes [51], aluminium hydroxide [52] or ion exchange resins [53]. The adsorbent is added to the water sample and the whole centrifuged and resuspended in a small volume. The concentration can be by a factor of 20 to 100,000 times.

Coxsackie virus B3 which had been seeded into 400 ml of tap water was recovered by flocculation with 200 mg of alum and subsequent elution of the virus from the floc by dissolving it at pH 9 in Tris buffer containing EDTA [54]. This technique has the advantage of being cheap and simple.

The technique which receives the most attention is membrane adsorption. Viruses can be adsorbed onto the matrix of membrane filters of the 'Millipore' or 'Gelman' type, even when the pore diameter of the filter is 10 to 20 times greater than the viral diameter. A suspension which contains the virus is passed through the membrane filter and the virus eluted with serum or gelatin [55] or sodium lauryl sulphate [56] or 3% beef extract [57]. The addition of 0.05M magnesium chloride will usually increase the adsorption. Alternatively the filter can be made of aluminium alginate gel [58] which after filtration can be dissolved in a 3.8% solution of sodium citrate. This technique was used to study the recovery of attenuated poliovirus 1 from up to 20 l of seeded distilled water and Thames River water. The maximum adsorption of virus occurred at pH 4 to 8.

The efficiency of recovery of poliovirus 1 from 380 l of seeded finished drinking water was studied using a wide variety of filters, including MF nitrocellulose membranes; AA Cox M-780; epoxy-fibreglass-asbestos discs; K-27 yarn-wound fibre glass cartridges plus AA Cox M-780 discs and Balston epoxy-fibreglass tubes [59]. They were all equally effective adsorbent filters for recovering the virus but the Balston filters had advantages in size, weight, cost and handling factors. Using the Balston filter input levels of 12-22 PFU/1,900 l could be detected. The sensitivity of a proposed standard method for water examination was studied with a wide range of filters [60]. The virus was absorbed at pH 3.5 and then eluted with glycine buffer at pH 11.5. From the 44 samples studied, poliovirus was detected at 95% reliability at a mean virus input level of 3-7 PFU/380 l when 1,900 l of water were sampled. When levels of < 1-2 PFU/380 l were considered in 76 samples the detection reliability was 66%. Members of the coxsackie virus groups A and B, echovirus and adenovirus were detected when 380 l or 1,900 l of water were sampled. The technique of membrane adsorption is sensitive and can detect low levels of virus in large volumes of water.

Other Indicators of Pollution

The extent of water pollution has also been assessed using diatoms as indicator organisms [61]. Diatoms constitute a group of unicellular algae with a relatively well known taxonomy and ecology. They can be useful, as can algae in general [62], for the monitoring of water pollution by organic and nitrogen compounds.

Oxygen Demand Tests

An assessment of pollution can be made using chemical methods, although several of these are based on estimates of microbial activity. The use of a chemical index in place of coliform counts has been suggested [63] and these techniques serve a useful purpose, but a detailed discussion of them is outside the scope of the present work.

Biological (Biochemical) Oxygen Demand (BOD)

This test is the most important and probably the most widely used when pollution is of the organic type. It simulates the actual conditions which are likely to be encountered when the effluent is discharged at a reasonable volume into a stream. Essentially the sample is contaminated with a rich bacterial flora and mixed at various dilutions with aerated distilled water and incubated at 20°C for five days. The concentration of dissolved oxygen is measured at the beginning and end of this period. The amount of oxygen consumed in this period is usually quoted as mg/l of effluent. The test depends on the presence of viable bacteria, and if these are not present in the test sample they must be added, e.g. as raw sewage.

The series of dilutions must be made so that one of them uses about half of the available oxygen, and this one gives the most reliable result. The dilutions used depend on the source of the sample, but approximations are as follows: settled sewage 1-5%; for oxidized effluents 5-25%; polluted river waters 25-100%; and for strong trade wastes 0.1-1% may be satisfactory. Because of the nature of the test there is no standard against which the accuracy of the BOD test can be measured. Reference standards can be used including glucose, which at 300 p.p.m. has a 5 day-20°C BOD of 224 p.p.m., or glutamic acid which has at 300 p.p.m. a 5 day-20°C BOD value of 217 p.p.m. However, glucose has a high and variable oxidation rate with simple inocula. It is therefore probably better to use a mixture of 150 mg/l of glutamic acid and 150 mg/l of glucose which has a

BOD similar to municipal waste. It is not always possible to replicate the test, and so it must be carried out for the exact time at the prescribed temperature.

In general the BOD test is useful for evaluating domestic sewage because this is a balanced medium for the growth of the microflora. It may be less reliable for industrial effluents, as the presence of toxic materials can interfere with the test. Average domestic waste has a BOD value of about 300 p.p.m., whilst the untreated effluent from a penicillin plant could be as high as 32,000 p.p.m. If these values are compared with rivers, it is found that very clean rivers have a BOD of less than 1 p.p.m., whilst rivers with a BOD of 10 p.p.m. or more are considered to show marked signs of pollution. Provided there is an adequate flow of water, effluent of 20 p.p.m. 5 day-20°C BOD can be introduced into a river if the suspended solids do not exceed 30 p.p.m.

If an effluent contains high numbers of <u>Nitrosomonas</u> and a high concentration of ammonia ion, then nitrification can interfere with the BOD test because of the high requirement for oxygen in the oxidation of ammonia to nitrite. These conditions occur in the incubation of partially nitrified sewage effluents. Nitrification can be suppressed, without interfering with carbonaceous oxidation, by incubating in the presence of allylthiourea at 0.5 mg/l [64]. This does not affect <u>Nitrobacter</u>, and so any nitrite present will still be oxidized.

The oxygen demand may also be obtained by a manometric method (Warburg apparatus) which can help to predict the amenability of an effluent to treatment by secondary processes.

Methylene-blue Stability Test

The methylene-blue test is a means of assessing the ability of an effluent to remain in an oxidized state when incubated out of contact with the air, i.e. its stability. The stability so determined is that of the biologically degradable material and also the initial concentrations of dissolved oxygen, nitrate and nitrite. The type and nature of the microbial flora will also influence the result.

The test is carried out in a bottle of 130 ml capacity to which is added 0.35 ml of a 0.05% w/v methylene-blue solution. The bottle is gradually filled with the sample and aeration is avoided as far as possible. The bottle is stoppered (without entrainment of air bubbles) and left at 20°C for five days. The contents are examined several times during the

first few hours, and then daily or twice daily. The time at which the blue
colour disappears is noted. If the blue colour persists after five days
the sample is considered stable and is recorded as having passed the
methylene-blue stability test. If the blue colour disappears before the
end of five days, the sample is relatively unstable and is recorded as
having failed the test in x days or hours.

Chemical Oxygen Demand Test (COD)

This test gives a measure of the total amount of organic matter in a
sample and does not distinguish between compounds which are biodegradable
and those which are not. Therefore, it is not a particularly good index of
effluent quality from any biological treatment plant where a significant
quantity of the organic content is not biodegradable. It is quite good for
assessing the strength of sewages or of trade wastes.

The test involves refluxing the sample with concentrated sulphuric
acid and a known amount of potassium dichromate, followed by titration of
the dichromate remaining when oxidation is complete. New improved versions
of the test have been reported in attempts to simplify the procedure
[65,66,67].

Permanganate Value

This is one of the oldest methods for assessing the strength of a
polluting liquid, and relies on the measurement of the oxygen absorbed
from acid-permanganate at a temperature a little above ambient. The
oxidation of organic material under these conditions is usually far from
complete and the measurements are often quite arbitary. If the sample contains materials which are easily oxidized e.g. sulphite or nitrite, then
false results may be obtained. The test does give a simple method of
assessing the quality of an effluent and can be used to estimate the
dilution required in the BOD test.

Microbiological Examination of Pharmaceutical Products

Micro-organisms may be pollutants and represent a hazard in both the
food and pharmaceutical industries. The situation for pharmaceutical
products is controlled, and limits are laid down in several pharmacopoeias
for certain raw materials or products. The food industry is not yet at
this formal stage, but the tests employed are similar and designed to

ensure that food products are free from food poisoning organisms.

Pharmaceutical products which have been packed under aseptic conditions are examined routinely for the presence of viable organisms. This testing for sterility is carried out to ensure that these products for injection or intravenous infusion do not contain any living organisms. The tests of the United States Pharmacopoeia XIX (USP) and the European Pharmacopoeia require that the preparation is tested for the presence of both aerobic and anaerobic bacteria and fungi.

In addition, certain pharmacopoeias have recognized that the presence of certain types of micro-organisms in raw materials, medicinal and drug products is undesirable. Both the British Pharmacopoeia (BP) 1973 [68] and the USP XIX [69] include tests for the presence of Escherichia coli and salmonellae, whilst the USP also specifies further tests for Staphylococcus aureus and Pseudomonas aeruginosa, and the BP specifies tests for pseudomonads.

The concern over non-sterile pharmaceutical products followed an out-break of salmonellosis which was considered to be due to thyroid tablets contaminated with Salmonella [70,71]. The level of microbial contamination can be determined in the raw materials and in manufactured oral and topical pharmaceutical products [72]. There may be high levels of contamination in some natural ingredients and additives such as starch, talc and water, which are used in the manufacture of pharmaceutical products.

The tests of the BP 1973 use an enrichment procedure for E. coli in which the raw material is added to 50 ml of nutrient broth and incubated at 37°C for 18-24 h. One ml of the resulting culture is added to 5 ml of MacConkey's broth. If acid and gas are produced then a secondary treatment is carried out using the Eijkman test and indole production at 44°C (see water testing). A positive control must be carried out using a culture of E. coli NCIB 9002 which contains 10-50 organisms/ml. Only if this gives a positive result can any result in the test be recorded. Tests which are not official may be applied to confirm the presence of E. coli, e.g. growth on eosin methylene-blue agar is suggested in the USP XIX. On this medium E. coli gives isolated colonies 2-3 mm in diameter which exhibit a greenish metallic sheen by reflected light and dark purple centres by transmitted light.

Eosin Methylene-blue Agar (Levine)

Peptone	10.0 g
Lactose	10.0 g
Dipotassium hydrogen phosphate	2.0 g
Eosin Y	0.4 g
Methylene-blue	0.065 g
Agar	15.0 g
Distilled water to	1 litre

The test for the presence of salmonellae requires a preliminary enrichment by placing the sample in 100 ml of nutrient broth and incubating at 37°C for 18-24 h. Following this, 1 ml of the culture is added to each of two tubes which contain (a) 10 ml of selenite F broth (see previously for details of medium) and (b) 10 ml of tetrathionate broth. These tubes are incubated at 37°C for 48 h.

Tetrathionate Broth

Beef extract	0.9 g
Peptone	4.5 g
Yeast extract	1.8 g
Sodium chloride	4.5 g
Calcium carbonate	25.0 g
Sodium thiosulphate	40.7 g
Distilled water to	1 litre

Mix, bring to the boil, cool to below 45°C, and add a solution of 6 g of iodine and 5 g of potassium iodide in 20 ml of water. Mix, and distribute in sterile containers.

The two cultures are then used to inoculate three plates which contain (a) brilliant green agar (b) desoxycholate citrate agar and (c) bismuth sulphite agar, and the plates are then incubated at 37°C for 18-24 h. The appearance of the colonies is characteristic of the salmonellae and is given in the table below.

Brilliant Green Agar

Peptone	10.0	g
Yeast extract	3.0	g
Lactose	10.0	g
Sucrose	10.0	g
Sodium chloride	5.0	g
Phenol red	0.08	g
Brilliant green	12.5	mg
Agar	12.0	g
Distilled water to	1 litre	

Mix, allow to stand for 15 min, sterilize by maintaining at 115°C for 30 min and mix before pouring.

Desoxycholate Citrate Agar

Beef extract	5.0	g
Peptone	5.0	g
Lactose	10.0	g
Sodium citrate	8.5	g
Sodium thiosulphate	5.4	g
Ferric citrate	1.0	g
Sodium desoxycholate	5.0	g
Neutral red	0.02	g
Agar	12.0	g
Distilled water to	1 litre	

Mix and allow to stand for 15 min. With continuous stirring, bring gently to the boil and maintain at boiling point until solution is complete. Cool to 50°C, mix, pour and cool rapidly. Care should be taken not to overheat desoxycholate citrate agar during preparation. It should not be remelted and the surface of the plates should be dried before use.

Bismuth Sulphite Agar

Dissolve the following with the aid of heat and sterilize by maintaining at 115°C for 30 min

(1) Beef extract   6.0 g
     Peptone   10.0 g
     Agar   24.0 g
     Ferric citrate   0.4 g
     Brilliant green   10.0 mg
     Distilled water to   1 litre

The following solution is now prepared:

(2) Bismuth ammonium citrate   3 g
     Sodium sulphite   10 g
     Anhydrous disodium hydrogen phosphate   5 g
     Dextrose   5 g
     Distilled water to   1 litre

Mix, heat to boiling, cool to room temperature, add 1 volume of solution
(2) to 10 volumes of solution (1) which has been previously melted and
cooled to a temperature of 55°C, and pour. Bismuth sulphite agar should be
stored at 2°-10°C for five days before use.

Medium Used	Appearance of Colony
Brilliant green agar	Small, transparent and colourless, or opaque, pinkish or white (often surrounded by a pink to red zone)
Desoxycholate citrate agar	Colourless and opaque, with or without black centres
Bismuth sulphite agar	Black or green

To differentiate between _Salmonella_ and _Proteus_ sp. the positive
colonies are subcultured, by surface inoculation and stab, in triple sugar
iron agar and into a tube of urea broth.

Triple Sugar Iron Agar

Beef extract	3.0 g
Yeast extract	3.0 g
Peptone	20.0 g
Lactose	10.0 g
Sucrose	10.0 g
Dextrose	1.0 g
Ferrous sulphate	0.2 g
Sodium chloride	5.0 g
Sodium thiosulphate	0.3 g
Phenol red	24.0 mg
Agar	13.0 g
Distilled water to	1 litre

Mix, allow to stand for 15 min, bring to and maintain at boiling point until solution is complete, mix, distribute in tubes and sterilize by maintaining at 115°C for 30 min. Allow to set in a sloped form with a butt about 2.5 cm long.

Urea Broth

Potassium dihydrogen phosphate	9.1 g
Anhydrous disodium hydrogen phosphate	9.5 g
Urea	20.0 g
Yeast extract	0.1 g
Phenol red	0.01 g
Distilled water to	1 litre

Mix. Sterilize by filtration and distribute aseptically in sterile containers.

The formation of acid and gas in stab culture (with or without blackening), and the absence of acidity from surface growth in triple sugar agar indicates Salmonella. The absence of red colour in urea broth is also indicative of Salmonella. If acid is produced without gas in the stab culture then serological methods must be used to confirm the identity of the organism. Positive control tests are carried out using a culture of

Salmonella abony NCTC 6017 which contains 10-50 organisms/ml.

The Enterobacteriaceae can be differentiated using commercial systems e.g. Enterotubes (Roche).

The test for the presence of Pseudomonas sp. is important because they are potential pathogens and they are resistant to many preservatives or can degrade them. They are also common contaminants of non-sterile pharmaceutical products [73] or environments [74]. The sample is placed in 100 ml of nutrient broth which contains 0.03% w/v cetrimide (a quaternary ammonium antiseptic) and incubated at 30°C for 72 h. After this the culture is subcultured onto solidified cetrimide medium and incubated for 48 h at 30°C. A comparative examination of the USP and BP tests for contamination of pharmaceutical products showed that of 28 strains of Pseudomonas, only 5 were detected by the BP method and 11 by the USP method. However, only one strain of Ps. aeruginosa was not detected using the USP method [75].

Any colonies produced on the cetrimide medium are examined microscopically by Gram's stain and Gram-negative rods recorded. The colonies may also produce a greenish pigment on cetrimide agar.

The oxidase test is also carried out, which consists of placing 2-3 drops of a freshly prepared 1% w/v solution of NNN'N'-tetramethyl-p-phenylenediamine hydrochloride on a piece of filter paper. This is smeared with a sample of the colony and if a purple colour is produced in 5-10 s the test is positive. The time is critical because other organisms may give a positive reaction in 2-5 min. Again, positive controls must be carried out using a suspension of Pseudomonas aeruginosa NCTC 6750.

The raw materials which have to be examined for microbial contamination are mainly drugs of natural origin, e.g. gelatin, digitalis leaf, pancreatin, thyroid, cochineal or tragacanth, where the requirement is that E. coli shall be absent from a 1 g sample and Salmonella from a 10 g sample. In addition a 1 g sample of Aluminium Hydroxide Gel BP must be free from Pseudomonas.

The bacterial content of tablets has been examined by several workers [76,77,78] and levels of over 100 bacteria/tablet have been found, but neither E. coli nor Salmonella were present. The viable numbers of bacteria present in the raw materials are reduced during the process of tabletting.

The technique of membrane filtration can be applied to the microbiological examination of soluble raw materials and may permit easier sampling of some materials. The membranes can be placed directly onto

selective media for identification purposes. Using this·technique combined
with enzymatic solubilisation several samples of gelatin were examined, and
E. coli was detected in some samples [79]. The technique may also be
applied to the examination of ointments and oils, if the oil is dissolved
in isopropylmyristate prior to filtration [80].

REFERENCES

1   D.J. Donsel and E.E. Geldreich, in L.J. Guarria and R.K. Ballantine
     (Editors), The Aquatic Environment: Microbial Transformations and
     Water Management Implications, EPA 430/G-73-008, Washington DC., 1973.
2   R. Feachem, Water Res., 8(1974)367.
3   World Health Organization, European Standards for Drinking Water, 2nd
     ed., WHO, Geneva, 1970.
4   World Health Organization, International Standards for Drinking Water,
     3rd ed., WHO, Geneva, 1971.
5   The Bacteriological Examination of Water Supplies. Reports on Public
     Health and Medical Subjects No. 71, H.M.S.O., London, 1969.
6   G.S. Wilson and A. Miles, Topley and Wilson's Principles of Bacteriology
     and Immunity, 5th ed., Edward Arnold, London.
7   F.S. Thatcher and D.S. Clark, Micro-organisms in Food, Vol. 1. Their
     Significance and Methods of Enumeration, University of Tc onto Press,
     Toronto, 1968.
8   E.F.W. Mackenzie, E.W. Taylor and W.E. Gilbert, J. Gen. Microbiol.,
     2(1948)197.
9   J.E. Jameson and N.W. Emberley, J. Gen. Microbiol., 15(1956)198.
10   R.D. Gray, J. Hyg., 62(1964)495.
11   E.H. Kampelmacher, A.B. Leussink and L.M. van Noorle Jansen :ater Res.,
     10(1976)285.
12   K.L. Thomas and A.M. Jones, J. Appl. Bact., 34(1971)717.
13   D.A.A. Mossel and C.L. Vega, Hlth. Lab. Sci., 10(1973)303.
14   G. Cavazzini, P. Cenci and L. Prati, Ig. Mod., 68(1975)167.
15   Public Health Laboratory Service Standing Committee on the Bacter-
     iological Examination of Water Supplies, J. Hyg., 66(1968)641.
16   D.A.A. Mossel, Lancet, (i)(1974)173.
17   V.A. Arbuzova, Trudy Inst. Epidem. Mikrobiol. Sanit., 36(1970)287.

18 W.H. Ewing, Differentiation of Enterobacteriaceae by Biochemical
    Reactions. U.S. Dept. of Health, Education and Welfare, CDC, Atlanta,
    1972.
19 J.E. Delaney, J.A. McCarthy and R.J. Grasso, Wat. Sewage Wks., 109
    (1962)289.
20 D.D. Mara, J. Hyg., 71(1973)783.
21 J.R. Matches, J. Liston and D. Curran, Appl. Microbiol., 28(1974)655.
22 O.R. Brown, Microbios, 7(1973)235.
23 B.J. Dutka, M.J. Jackson and J.B. Bell, Appl. Microbiol., 28(1974)474.
24 J.M. Anderson and A.C. Baird-Parker, J. Appl. Bact., 39(1975)111.
25 M.H. Brodsky and D.A. Schiemann, Appl. Environ. Microbiol., 31(1976)695.
26 R.L. Cada, Appl. Microbiol., 29(1975)255.
27 S.D. Lin, Ill. State Water Surv. Rep. Invest., 78(1974)1.
28 E. Hammerström and V. Ljutov, Acta Path. Microbiol. Scand., 35(1954)365.
29 V. Ljutov, Acta Path. Microbiol. Scand., 35(1954)370.
30 D.J. Reasoner and E.E. Geldreich, Proc. 2nd Amer. Water Works Assn.
    Tech. Conf., Dallas, Texas, 1974.
31 J.R. Schrot, W.C. Hess and G.V. Levin, Appl. Microbiol., 26(1973)867.
32 U. Bachrach and Z. Bachrach, Appl. Microbiol., 28(1974)169.
33 E.U. Buddemeyer, Appl. Microbiol., 28(1974)177.
34 P. Khanna, Water Res., 8(1974)311.
35 J.R. Wilkins, G.E. Stoner and E.H. Boykin, Appl. Microbiol., 27(1974)
    949.
36 J.S. Newman and R.T. O'Brien, Appl. Microbiol., 30(1975)584.
37 A.P. Pugsley and L.M. Evison, J. Appl. Bact., 38(1975)63.
38 J.E. Benbough and K.L. Martin, J. Appl. Bact., 41(1976)47.
39 J.L. Melnick, in V. Snoeyinck (Editor), Proc. 13th Water Quality
    Conference. Virus and Water Quality: Occurrence and Control, 15th-16th
    February, 1971, University of Illinois College of Engineering, Urbana,
    Ill., p. 114.
40 J.L. Melnick, J. Emmons, E.M. Opton and J.H. Coffey, Amer. J. Hyg.,
    59(1954)185.
41 B.D. Rawall and S.H. Godbole, Environ. Health, 6(1964)234.
42 G. Berg, D. Berman, S.L. Chang and N.A. Clarke, Amer. J. Epidemiol.,
    83(1966)196.
43 N.G. Anderson, G.B. Cliver, W.W. Harris and J.G. Green, in G. Berg
    (Editor), Transmission of Viruses by the Water Route, Wiley-Interscience,
    New York, 1967, p. 75.

44 M. Bier, G.C. Bruckner, F.C. Casper and H.B. Roy, in G. Berg (Editor), Transmission of Viruses by the Water Route, Wiley-Interscience, New York, 1967, p. 57.

45 T. Gibbs and D.O. Cliver, Hlth. Lab. Sci., 2(1965)81.

46 D.O. Cliver, in G. Berg (Editor), Transmission of Viruses by the Water Route, Wiley-Interscience, New York, 1967.

47 P.A. Albertsson, Fractionation of Cell Particles and Macromolecules in Aqueous Two-phase Systems, Almquist and Wilksells, Uppsala, 1960.

48 G. Frick and P.A. Albertsson, Nature, 183(1959)1070.

49 H. Shuval, in Developments in Water Quality Research, Ann Arbor-Humphrey Science Publishers, Ann Arbor, 1970.

50 C. Wallis and J.L. Melnick, Am. J. Epidemiol., 85(1966)459.

51 S. Grimstein, J.L. Melnick and C. Wallis, Bull. World Hlth. Org., 42(1970)291.

52 J. Taverne, I.H. Marshall and R. Fulton, J. Gen. Microbiol., 19(1957) 451.

53 C.R. Gravelle and T.D.Y. Chin, J. Infect. Dis., 109(1967)205.

54 S.M. Lal and E. Lund, in Advances in Water Pollution Research, Proc. 7th Int. Conf., Paris, Sept. 1974, 1975.

55 D.O. Cliver, Appl. Microbiol., 13(1965)417.

56 C. Wallis and J.L. Melnick, J. Virol., 1(1967)472.

57 H.A. Fields and T.G. Metcalf, Water Res., 9(1975)357.

58 H. Gastner, in G. Berg (Editor), Transmission of Viruses by the Water Route, Wiley-Interscience, New York, 1967, p. 36.

59 W. Jakubowski, W.F. Hill and N.A. Clarke, Appl. Microbiol., 30(1975)58.

60 W.F. Hill, W. Jakubowski, E.W. Akin and N.A. Clarke, Appl. Environ. Microbiol., 31(1976)254.

61 H. van Dam, Hydrobiol. Bull., 8(1974)274.

62 S.L. Vanlandingham, Hydrobiologia, 48(1976)145.

63 D.L. Lamba, P. Singha and B.L. Sharma, Indian J. Med. Res., 62(1974) 1808.

64 L.B. Wood and H. Morris, J. Proc. Inst. Sew. Purif., (1966)350.

65 E. Canelli, D.G. Mitchell and R.W. Pause, Water Res., 10(1976)351.

66 S.M. Dhabadgoankar and W.M. Deshpaude, Indian J. Environ. Health, 16(1974)300.

67 A. Forsberg and S.O. Ryding, Vatten, 31(1975)148.

68 British Pharmacopoeia, London, H.M.S.O., 1973, p. A121.

69 United States Pharmacopoeia XIX revision, Mack Publishing Co., Easton, Pennsylvania, 1975, p. 588.

70 L.O. Kallings, F. Ernerfeldt and L. Silverstolpe, The 1964 Inquiry by The Royal Swedish Medical Board into Microbial Contamination of Medical Preparations: Final Report, 1965.

71 L.O. Kallings, O. Ringertz, L. Silverstolpe and F. Ernerfeldt, Acta Pharm. Suecica, 3(1966)219.

72 G. Sykes, J. Mond. Pharm., 14(1971)8.

73 E.G. Beveridge and I.A. Hope, Pharm. J., 207(1971)102.

74 Anon, Br. Med. J., (i)(1976)958.

75 A. Hart, K.E. Moore and D. Tall, J. Appl. Bact., 41(1976)235.

76 A. Fischer, B. Fuglsang-Smidt and K. Ulrich, Dansk. Tidss. Farm., 42(1968)125.

77 B. Fuglsang-Smidt and N. Ulrich, Dansk. Tidss. Farm., 42(1968)295.

78 R. Gallien, Pharm. Ind., 34(1972)929.

79 K.A.C. Chesworth, A. Sinclair, R.J. Stretton and W.P. Hayes, J. Pharm. Pharmacol., 29(1977)60.

80 A. Hart and M.B. Ratinski, J. Pharm. Pharmacol., 27(1975)142.

SEWAGE TREATMENT

Sewage is the liquid waste of a community and it comes from three
main sources: domestic sewage which carries human excrement and a variety
of detergents; industrial sewage which includes a wide variety of organic
and inorganic chemicals and may contain pathogens if it comes from a
slaughterhouse; and storm sewage, which may contain any pollutants washed
off in the rainwater. Of these, only storm sewage will be run off into a
body of natural water.

Farm effluents are much stronger, more difficult and expensive to
treat than domestic sewage. They can be dealt with alone, or mixed with
domestic sewage, at a local authority works. The waste should first be
treated by settlement and the tanks regularly cleaned out. Wherever
possible, the waste should be used in agriculture because it contains
valuable plant nutrients.

The treatment of the waterborne wastes is designed to produce water
of a definite quality. The quality required is governed by the use which
is ultimately going to be made of the wastewater. This water may be
reused as a raw water source for industry, or for a town water supply, or
it could be used directly for recreational purposes.

Treatment of wastewater can take place as soon as the waste is
collected and before discharge. Alternatively, natural purification in
the surface and groundwater can be used, or the water can be processed for
use in the production of drinking or industrial water. These processes
are often arbitarily separated, but they are really one integrated process,
each removing some of the waste materials. The proper treatment of wastewater is of great and increasing importance, since the raw water supply of
a growing number of cities is supplemented with the sewage effluent of an
upstream town or city. For example, most of London draws its water from
the lower River Thames and the River Lea, both of which receive wastewaters from towns upstream.

The use of rivers as cheap aqueducts to provide a raw water supply
started several generations ago, and relied (from the latter part of the
19th century) on the ability of filtration and chlorination to remove the
organisms responsible for typhoid, cholera and dysentery. These treatments
may be inadequate to deal with viruses and the chemicals being used today
as pesticides and herbicides. The disinfection treatment carried out at

one point may present a pollution problem downstream, e.g. trihalomethanes and chloroform may be present in finished waters after treatment with chlorine [1].

The objective of water treatment is to reduce the amount of any pollutant to a safe level, which is determined by the use to be made of the water. The water may be rigidly defined for drinking water or for industrial use. One important factor in determining a safe level is the cost involved in removing a particular pollutant. The problem is complicated by the increasing number of chemicals in use whose long term effects are not known. The treatment of water is certainly effective in reducing the spread of waterborne bacterial diseases, and the processing required to achieve this is relatively cheap. Consideration can be made, on a mathematical basis, of waste load and its effect on water quality and the options available for improving water quality can be calculated [2,3].

Micro-organisms have a key role in all the stages of waste treatment and the processes are based on an intensive, controlled mineralisation by aerobic processes (activated sludge, trickling filters or slow sand filters) or by anaerobic digestion (methane fermentation). Both processes give products which are harmless and inoffensive (CO_2, H_2O) and easily removable (cells, CH_4). Any pathogenic organisms are removed because they are used as additional organic material for mineralisation.

Settled sewage is like a diluted culture medium but it has a high requirement for dissolved oxygen. The low solubility of oxygen is of great importance in sewage treatment, because if it is used up the sewage will rapidly become anaerobic. Therefore, there is a requirement in all the processes for adequate aeration by stirring, and for the estimation of the amount of oxygen required for the microbial oxidation of the organic material present in the sewage (BOD). The amounts of nitrogen or phosphate in a sewage are rarely limiting, but the problem with these nutrients occurs when the effluent is discharged, because then it is an ideal medium for algal growth (see chapter on eutrophication).

If a settled sewage is left to degrade by anaerobic processes volatile reaction products occur, e.g. H_2S and NH_3 which are supplemented with the breakdown products of fatty acids and proteins. These combinations produce very objectionable odours and so any processes involving this require covering. These odours are very much reduced when the waste is adequately oxygenated, and this principle is utilized in the activated sludge and trickling filter processes.

114

Fig. 1.   Diagrammatic Representation of Sewage Treatment Processes.

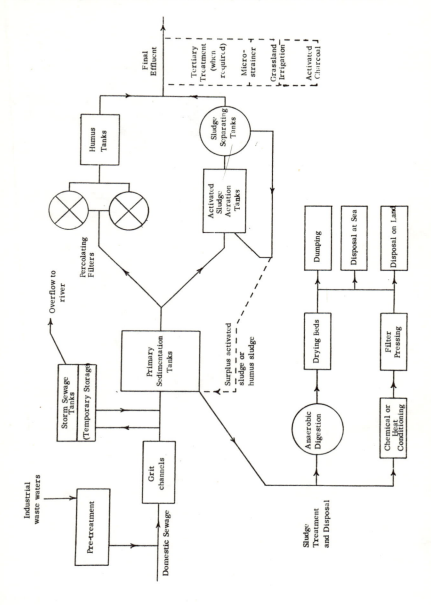

The processes used for the treatment of sewage are outlined in Fig. 1. The primary treatment of sewage consists of the removal of solid material by physical means. Large solid objects can be removed using metal bar screens and grit can be deposited by slowing the flow rate to 0.3 m/s. The sewage can then be fed to tanks where it may be held for up to 15 h. Stronger sewages are more amenable to settling for removal of solids than are weaker sewages. An efficient tank should remove about 60% of the suspended solids and about 40% of the organic material. It is important to remove the putrefactive raw sewage sludge, which is formed by compaction of the settled solids. If this material is allowed to accumulate, anaerobic digestion will start and the methane produced bubbles up. This will cause sludge particles to be held in suspension because of the agitation and the efficiency of sedimentation will be reduced.

The methods of treating the settled sewage include the use of activated sludge, biological filtration, oxidation ponds and land treatment.

Activated Sludge Process

The activated sludge process is similar to a continuous culture system, but it differs because there is a continuous inoculation of the aerated culture with micro-organisms present in the incoming sewage and in the feedback of cells (as activated sludge) to the culture vessel. These features give a rapid adsorption, uptake and oxidation of pollutants. They also help to maintain a stability of operation over a wide range of incoming nutrient concentrations and dilution rates. The main advantage it has over other treatments is its compactness for a given rate or removal of substrate.

As in any sewage treatment, the concentration of nutrients and cells in the final effluent must be low. The conditions in the aerating tank must give compact flocculent growth so that the settling is rapid. If filamentous micro-organisms grow a "bulking sludge" is produced which results in hindered settling and the escape of organisms in the effluent.

The design of plant for the activated sludge process has been dealt with by Hawkes [4], Bolton and Klein [5], Coackley [6] and Abson and Todhunter [7].

The initial stages of aeration give the most rapid oxygen uptake, and this is often provided by having a greater degree of aeration at the inflow end of the tank (tapered aeration), or by dividing the inflow and introducing it at intervals along the aeration tank (step aeration).

Fig. 2.　　Typical Examples of Protozoans Found in Wastewater,
Particularly Activated Sludge.

Vorticella sp.

Carchesium sp.

Euplotes sp.

Opercularia sp.

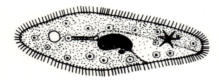

Paramecium sp.

Trachelophyllum sp.

Aeration may be achieved by supplying diffused air or by vigorous mechanical aeration, and both processes will mix the activated sludge liquid. The mixing is usually not complete and the conditions will depart from the ideal [8] which is a homogenous completely mixed reactor with feedback. The real situation is more like a tubular, pipe flow reactor.

i) Ecology of Activated Sludge

The microbial population of the activated sludge process is specialized and has a relatively low diversity of species. Heterotrophic bacteria dominate and with the saprobic (feeding on dead material) protozoa constitute the basic trophic level. Then come the holozoic protozoa feeding on bacteria. Fungi are present in low numbers, except in certain bulking conditions. There may also be low numbers of rotifers and nematode worms, but algae are normally absent. The ecology of the activated sludge process has been discussed in detail by Hawkes [4] and Pipes [9].

The numbers of bacteria present in activated sludge are difficult to determine until the flocs have been broken up [10]. The viable count is found to be one or two orders of magnitude greater when the flocs have been broken down by homogenization. The problem of examining the bacterial flora is complicated because there is no single medium which will support the growth of all the nutritional types of bacteria present in an activated sludge [11].

The majority of bacteria are Gram-negative organisms and the coli-aerogenes group account for only a small fraction of the population [12]. The level of E. coli in the total coli-aerogenes population falls from 75% in raw sewage to 25% in activated sludge, and to 30% in effluent [13]. Nitrosomonas sp. are present and can be isolated by an enrichment technique [14] or their presence detected by chemical analysis of the liquid.

The heterotrophic bacteria in activated sludge were generally unaffected by the addition of alum, for phosphate precipitation. However, a lower number of nitrate oxidizing bacteria and a reduced frequency of protozoa types was noted when compared to sludges which had not been dosed with alum [15].

The bacterial population may vary depending on the type of waste being treated. When antibiotic wastes were considered 85 strains, 4 genera and 16 species of bacteria were identified and Pseudomonas was dominant [16].

Protozoa are present in large numbers and considerable variety in activated sludge, and those found most frequently are shown in Fig. 2 .

The ciliates are protozoa which have many cilia (hair-like extensions) and
may reproduce asexually by transverse fission, whilst flagellates possess
one or few whip-like extensions (flagella) and reproduce asexually by
longitudinal fission. About 228 species of protozoa have been reported in
activated sludge and the ciliates usually predominate. When there are few
ciliates and many flagellates present the sludge is probably overloaded
with organic material and in poor condition. A consideration of all the
species of protozoa present in a sludge allows the approximate BOD range
of the effluent to be predicted from the species structure [17]. When a
survey of 36 sewage treatment plants was carried out, predictions which
used the protozoan population of the sludge were correct in 80% of the
cases. The use of this type of indicator system may be misleading because
it oversimplifies a complex ecosystem.

The role of protozoa in the activated sludge process has been the
subject of speculation. Fairbrother and Renshaw [18] suggested that they
were harmful and should be removed. However, opinions have changed since
then and they are now regarded as being beneficial. The importance of
protozoa in producing a good quality effluent was demonstrated by using
protozoa-free activated sludges obtained from sludge taken from a full-
scale plant [19]. The mixed bacterial population in the sludge was kept
free from protozoa by adding cooled, heat-treated sewage. Under protozoa-
free conditions the sludge produced a very turbid effluent which had a
high five day BOD and the level of organic carbon and suspended solids was
high. The turbidity was directly related to the very high numbers of
bacteria which were suspended in the effluent. If protozoa were added and
the population was allowed to stabilize, the quality of the effluent
improved.

The protozoa may also play a part in flocculation by secreting mucous,
from the peristome in Balantiophorus minutus [20]or a mucous and poly-
saccharide in Paramecium caudatum [21].

Fungi, and to a lesser extent yeasts, can be isolated from activated
sludge [22], but they are not a significant or important part of the flora.
About 20 species of fungi have been isolated from samples taken from activated sludge plants [22]. The highest proportions of the colonies
observed were of Cephalosporium (38%), Cladosporium cladosporoides (22%)
and Penicillium sp. (19%), and yeasts accounted for only 1% of the colonies.
Similar results were obtained from activated sludge in Auburn, Alabama
[23]. Fungal numbers fall to their lowest level in winter and rise to a

peak in the autumn. Predacious fungi e.g. Zoophagus sp. and Arthrobotrys
sp. have been found in pilot plants where they captured rotifers and
nematodes [24].

The 'bulking' of sludge can be caused by Geotrichum sp. which may
dominate the flora of a plant [25]. When nitrogen or phosphate are
limiting Geotrichum may have a competitive advantage over the rest of the
microbial population [26].

ii) Flocculation

Flocculent growth is required for the production of a clear effluent
and for adequate concentration of the sludge, which is returned from the
gravitational settling tank. The flocculent growth is not required for the
removal of substrate material.

It was considered that a single bacterium, Zooglea ramigera, was
responsible for flocculation because it secreted a gelatinous material, to
which other organisms adhered and were enveloped. However, it has been
shown that a wide variety of bacteria can be isolated, from activated
sludge flocs, which are able to flocculate in aerated pure cultures [27].
If a plant is operated at a high dilution rate and high substrate concentrations then a more dispersed, filamentous type of growth is produced.
If the opposite conditions prevail then flocculation is obtained [28].

There are many theories which attempt to explain flocculation. It
may be caused by polyelectrolytes e.g. polysaccharides and polyamino acids,
which are excreted during the endogenous or decline phases of growth [29].
This approach suggests that bacterial suspensions resemble protected
dispersoids and experimental flocculation can occur if cationic polyelectrolytes are present in the treatment liquor. These polyelectrolytes
can be produced as a result of metabolism, or may occur naturally, e.g.
humic acid [30].

There is also evidence to suggest that the production of poly-β-
hydroxybutyric acid is associated with floc production [31]. However,
there is possibly more evidence which suggests that there is no correlation
between poly-β-hydroxybutyric acid level and floc production [32,33].

Flocculation can be brought about by low pH values. The presence of
calcium and magnesium ions can either produce flocculation [34] or reverse
it [35]. The flocculation of bacteria is also independent of the C:N ratio
[34].

iii) Bulking of Sludge

The bulking of sludge interferes with plant operation and effluent quality.

When a bulking sludge is produced it has very poor settling qualities because of the flocculent, filamentous growth of micro-organisms which was regarded as being due solely to bacteria of Sphaerotilus sp. It is probably due to a population which may be dominated by Sphaerotilus sp. but other filamentous bacteria belonging to genera such as: Bacillus, Nocardia, Beggiatoa, Thiothrix, Vitreoscillaceae, Leucothrix, Lineola and the fungal genera: Geotrichum and Zoophagus may be present.

Sphaerotilus sp. can utilize a variety of carbohydrates, and either amino nitrogen or inorganic nitrogen provided vitamin B_{12} is present. Bulking is encouraged by a low dissolved oxygen concentration, the combined effect of high C:N and C:P ratios [36], or N deficiency [37].

When the related 'sewage fungus' grows in polluted water it may constitute a nuisance [38]. Sewage fungus is a sensitive indicator of organic pollution in flowing waters and it can be a serious problem. If conditions are suitable it gives rise to a massive growth of cotton-wool like plumes which rapidly colonise all submerged surfaces.

The sewage fungus is a matrix of filamentous organisms which hold the whole community together. The organisms present [38] are the bacteria: S. natans, Zooglea sp., Flavobacterium sp. and Beggiotoa alba, the fungi: Geotrichum candidum, Leptomitus lacteus and the alga: Stigeoclonium tenue. Also present are motile, non-filamentous organisms (protozoa, diatoms and bacteria). Higher organisms such as nematodes and rotifers may be present, but will be feeding on the slime.

Sewage fungus occurs in rivers below sources of organic pollution. This may be treated or untreated domestic sewage, or untreated effluent from the food and drink industries. The growth is usually found in a zone just below the effluent discharge point, where the carbon content is high. In this region the algal numbers are low, but further downstream there may be a region where the algal numbers are high and may reach the level of a bloom.

iv) Effectiveness of the Activated Sludge Process

The process will remove up to 95% of the five day BOD, 95% of the suspended solids and up to 98% of bacteria and viruses from settled sewage.

The activated sludge process is effective in reducing the numbers of coliform organisms, Salmonella, Shigella and Mycobacterium tuberculosis by 85 to 99%, and polio virus type I by 90% and coxsackie A9 virus by 98%. The reduction in level of enteric viruses is aided by adsorption onto flocs. The process does not have a great effect on the cysts of parasitic protozoa and worm ova.

A significant reduction in the numbers of bacteria takes place as a result of ingestion by protozoa, and this predation may account for the removal of dispersed bacteria without considering flocculation [39,40]. Reductions of 90 to 99% have been recorded by this means for E. coli, Salmonella sp., Shigella sp. and Vibrio cholerae. Even when bacteriophage and Bdellovibrio sp. are present, which are specific for enterobacteria and pseudomonads, they are not considered to be effective in removing bacteria [13].

Heavy metals can be removed by the activated sludge and up to 78% of Pb is removed, but only 1% of the Ni. Other metals fall between these extremes [41].

If B. cereus is present then it can metabolize up to 1,000 mg/l of phenol [42] (see chapter on aromatic degradation). The system is capable of operating in the presence of toxic materials. When picric acid and lindane were used as model industrial toxicants, the catalase and dehydrogenase activity of the activated sludge dropped sharply and then recovered up to concentrations of 200 mg/l [43]. Also, thiocyanate carbon can be metabolized to CO_2 by activated sludge cultures and enzymes of the tricarboxylic acid cycle were not altered under experimental conditions [44].

Oxidation Ditch

There are many variations of the activated sludge process, and one of the more important is the oxidation ditch. In this the dissolved and suspended organic material are treated simultaneously in an ellipsoidal ditch which is provided with brush aeration. The ditch may simply be used as an aeration tank, with the brushes on; or as a sedimentation tank, with the brushes off, and it will operate as a stirred reactor with feedback of cells. Instead of a few hours the detention time will be between one and three days, depending on the temperature. This longer detention time leads to the production of a small amount of sludge which can be dewatered, on drying beds, without the risk of producing offensive odours. If full nitrification is required it can be obtained by stopping the rotors, when nitrate reduction and anaerobiosis commence.

Biological Filtration (Trickling Filter)

This type of filter has been in use for some seventy years in the United Kingdom and has been extensively studied. Its performance depends on careful attention to the design of plant, to maintenance and to method of operation [45,46].

The settled sewage is distributed on the surface of a bed of inert material which is 4 to 6 feet thick. After the filter has been in operation for a short time, a gelatinous, zoogleal slime of biological growth develops. The size of the inert material is large enough to permit air and water to pass over the organisms. Oxygen is absorbed from the air by the falling film of wastewater. The natural draught which is created causes air to flow upward through the filter at the same time as the waste-water passes down.

There is a stratification of organisms in the filter, because the sewage reaching the lower depths contains less organic material than the applied sewage. This favours the development of an autotrophic community at the lower levels.

Many industrial effluents which contain toxic materials, or small concentrations of organic material of natural origin, can be treated by biological filtration. This process seems better able to withstand shock loads of toxic material than the activated sludge process. Thus, phenolic compounds can be oxidized to CO_2 and water by Vibrio sp. and Pseudomonas sp., thiosulphate and thiocyanate can be oxidized by the thiobacilli and cyanide is ultimately metabolized as thiocyanate. At the cleaner, lower depths of the filter Nitrosomonas and Nitrobacter predominate. Fungi occur in greater number and variety than in the activated sludge process.

The excess biological growth is removed by the flushing action of the sewage and by the grazing action of fly larvae and worms. The dead film and larval faeces are separated from the filter effluent by a short period of sedimentation. The amount of humus discharged depends on the season, being particularly large during the spring. If excess humus is produced then ponding can occur. The humus sludge can be returned to the incoming sewage and removed by primary settling.

Filters require to be matured before they are efficient and, in the United Kingdom, seem to mature best if started in the spring or summer. If a filter is started too late in the year it may pond during the winter because an active population of flies, springtails or other scouring organisms has not become established. It requires three to six months for a filter to mature.

The biological filter removes up to 95% of the five day BOD, suspended solids and bacteria. The level of Salmonella paratyphi B is reduced 84 to 99%, Mycobacterium tuberculosis by 66%, total coliforms by 85 to 99%, enteric virus by 40 to 60%, tapeworm ova by 18 to 70% and cysts of Entamoeba histolytica by 88 to 99%.

Rotary-disc Biological Contactors

Purification plants using this process have only recently been introduced into the United Kingdom. They have been in use on the Continent, particularly in Germany, for some time, mainly for the treatment of sewage from small communities. The design and operation of plant for this process has been reviewed by Antonie [47].

Treatment begins with conventional primary settlement of the crude sewage and the settled sewage passes on to a biological stage. This is a concrete tank, or series of tanks, in which an assembly of closely-spaced vertical plastic discs (1-3 m diameter) is continuously rotated on a single horizontal axle which is motor-driven at 1-3 rev./min. About 40% of the surface area of the disc is submerged in the wastewater. Like the medium in a biological filter the discs serve as a support for the growth of a microbial film. The organisms present in the wastewater adhere to the support and build up a microbial film of up to 4 mm in thickness. As it rotates the contactor carries a film of wastewater into the air, which trickles down the surfaces and oxygen is absorbed.

The organisms present in the film remove organic material and consume oxygen. The excess microbial film is removed by a shearing action as the contactor passes through the water and the continuous wetting of the film prevents the development of filter flies.

After the contact with the discs the effluent passes through a secondary treatment settlement tank for the removal of the humus which has been detached from the discs.

Oxidation Ponds

A comprehensive method of treating whole waste is by using an oxidation pond [48]. These have an important place as simple and cheap devices in regions where adequate sunlight is available and land is inexpensive. The theoretical basis for their operation is not well formulated and there are many variations of the system.

The waste is discharged into a pond 3-5' deep and mixed with an algal and bacterial culture. The bacteria oxidize organic material to CO_2, H_2O, NH_3 and other decomposition products. These in turn are used as nutrients by the algae. The oxygen produced photosynthetically by the algae is used by the bacteria. An alternative would be if the organic material was converted to methane by anaerobic processes and the algal growth served as a cover to hold in the offensive odours.

The actual process is probably somewhere between the two, as some anaerobic digestion will take place. The change in BOD is probably small, because the waste is transformed into other organic material as bacterial and algal cells. This is not a problem, provided the algal mass can be harvested and used economically. The treatment of wastes in closed environments is possible using this process and, if the algae are digested under anaerobic conditions, the methane produced can be used as a fuel i.e. a conversion of light energy to chemical energy.

Land Treatment

In this process which is obsolete in the United Kingdom, settled sewage percolates intermittently over arable land. The accumulated organic material is removed by microbiological oxidation. This method of treatment is now confined to the polishing of effluents from an activated sludge or biological filtration process.

Where raw sewage is used as a fertilizer then there are problems with the transmission of human pathogens, particularly of protozoa and helminths. This may be a particular problem in the Far East.

Sludge Treatment

Sludge is formed at several points in the treatment plant and its disposal can present problems. The scale of the problem is shown by the fact that in the United Kingdom over 10^6 tons dry weight of sludge are produced each year at local authority plants alone.

The sludge obtained from the screening site and the sedimentation tanks is highly putrefactive and objectionable. Therefore, rapid disposal is essential. This sludge contains only 5-8% of solids and it is difficult to remove the water. The liquid sludge may be disposed of by dumping at sea or by discharge into permanent lagoons, but it is usually subjected to further processing.

In most regions the sludge cannot be disposed of as a liquid, but it

must be dried to about 25% solids. This produces a cake which can be
handled as a solid. The drying processes can be quite varied and include
heat treatment, filtration and rotoplug concentration. The most commonly
used method in the United Kingdom is probably dewatering on drying beds,
even though this method is affected by the weather. Usually, it is only
during the summer months that the sludge becomes dry enough to lift, and
so there can be heavy demands on labour for short periods of the year. The
wet sludge has to be stored, either on the beds, or in tanks. The concentration of heavy metals is increased in the soil beneath the sludge holding
ponds and the metals move as the soluble organo-metalic complexes.

The mechanical methods for dewatering, such as filter presses are
increasing in use. These methods need much less land than do drying beds
and are less dependent on the weather. For economic operation of filter
presses, the sludge has to be conditioned by treatment with agents which
cause aggregation of the sludge particles. An advantage of using filter
presses is the reduction of the nuisance from offensive odours.

i)   Digestion of Sludge

The sludge can be subjected to anaerobic digestion, which, in the
United Kingdom, is the most widely used single process. Anaerobic
digestion differs from putrefaction in that no offensive odours are
produced. This process could also be applied to strong organic industrial
wastes, or from farms. It has been applied to the digestion of wastes
from slaughterhouses. The methane produced as a result of the process can
be used as an energy source. Certain organic materials like lignin are
digested so slowly that they are for all practical purposes non-digestible.

A wide variety of bacteria are involved in the process of anaerobic
digestion, but they can be classified into two broad groups, the acid
producers (non-methanogenic bacteria) and methane producers (methanogenic
bacteria). The acid-producing bacteria will degrade most organic materials
(whether they are in solution or suspension) to give mainly lower fatty
acids; acetic acid accounts for about 80% of the total. The lower
aldehydes and ketones are produced in much smaller amounts. If cysteine
is taken as an example the reaction can be expressed as:

$$H_2N-\underset{\underset{CH_2SH}{|}}{\overset{\overset{COOH}{|}}{CH}} + 2H_2O \longrightarrow CH_3COOH + CO_2 + NH_3 + H_2S + 2H$$

The methanogenic bacteria convert the soluble materials which are produced by the acid-producers into a mixture of methane and carbon dioxide:

$$4CH_3COOH + 8H \longrightarrow 5CH_4 + 3CO_2 + 2H_2O$$

The acid-producers have a short doubling time, which can be measured in minutes or hours. Therefore, they will quickly become dominant when an organic waste is stored and so cause putrefaction. The methanogenic bacteria, on the other hand, have a longer doubling time e.g. four to six days and are easily inhibited by fatty acids, ammonia or soluble sulphide. The conditions produced during growth of the acid-producers therefore favour putrefaction, and these organisms must be controlled to permit development of methanogenic bacteria. The two groups have to develop in a balanced way.

The amount and composition of digester gas which is produced by complete digestion depends on the composition of the organic substrate and can be calculated [49]:

$$C_nH_aO_b + (n - a/4 - b/2)H_2O = (n/2 - a/8 + b/4)CO_2 + (n/2 + a/8 - b/4)CH_4$$

When $n > a/4 + b/2$, water is utilized in the reaction and the weight of gas exceeds the dry weight of organic material degraded.

The number of bacteria produced anaerobically is much less than is produced during aerobic growth. The anaerobic digestion process requires bacterial nutrients, the most important being nitrogen and phosphate. The minimum requirement for nitrogen is considered to be about 25% of the organic carbon and the phosphate about 20% of that for nitrogen.

The presence of ammonia, which may occur in farm wastes, is toxic to bacteria. In the United Kingdom anaerobic digestion is usually only applied to sewage sludge, but it could be more widely used in the treatment of pig effluent or poultry waste.

The methane which is produced biologically is always mixed with carbon dioxide, so the calorific value of digester gas will always be less than that of methane. The amount of gas produced from the digestion of sewage sludge is about 0.028 m^3/person/day in the United Kingdom. The thermal value of this gas represents only about 0.3% of the annual consumption of coal. Even if the waste from piggeries was digested, this could produce only 7% more gas than the human wastes.

The gas produced can be utilized, provided proper precautions against fire and explosion are taken, by burning for heat or to operate a stationary, modified petrol- or diesel-driven generator. The engine efficiency for generating electricity is probably only about 25%, but the heat generated can be recovered for heating the digester.

The gas could be used for vehicle propulsion, but this is a less attractive proposition [50]. The gas must be washed to remove the CO_2 and then it needs compressing to near the legal maximum of 20,000 kN/m^3 (3,000 lb/in^2). A 4.8 m^3 (170 ft^3) cylinder of methane, which would weigh about 68 kg (150 lb), will give the propulsion equivalent of 4.5 l (1 gal) of petrol. Also, should the gas be used to drive the compressor then a significant proportion would be taken for this.

The process of anaerobic digestion allows over 90% of the grease content to be turned into gaseous products (grease accounts for up to 20% of the dry solids of raw mixed sludge). Pathogenic organisms are reduced in viable numbers and the eggs of _Taenia saginata_ and cysts of _Heterodera rostochiensis_ are killed. Viruses which are present in the wastewater become bound to solids and are transported to the digester. Polio virus is fully recoverable from the sludge, but its infectivity is decreased in proportion to the time and temperature of the sludge digestion [51,52].

ii) _Agricultural Use_

In 1970 it was found that about 40% of the total sludge produced at inland sewage works in England and Wales was applied to agricultural land. In view of unsatisfactory results obtained in the past, the practice has been viewed with suspicion. With improved procedures the application of sewage sludge could meet part of the agricultural requirement for nitrogen. The sludge is cheap, but is deficient in potash and phosphate, and so is not a perfect substitute for farmyard or artificial fertilizers.

There may also be a health hazard from using sewage sludge as a fertilizer because of the possibility of an accumulation in plants of trace metals e.g. zinc, copper, nickel, cadmium, mercury and lead [53]. The concentration of heavy metals available for absorption by the plant depends on the pH of the soil. The heavy-metal content of food grown on land fertilized with sewage sludge has been studied [53,54]. Other toxic materials, such as boron from packaged detergents, can also occur in sludge.

Fig. 3.   NITROGEN CYCLE SIMPLIFIED

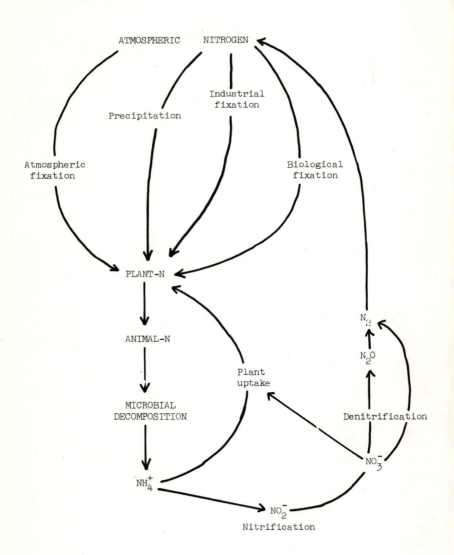

If no restrictions are placed on the amount of sewage sludge put onto the land, there may be a nitrate pollution problem from the loading of nitrogenous material [55].

The sludge can be dried, used as a low grade fuel and reduced to an inert and easily disposable ash. The difficulties associated with this are those concerned with drying the sludge adequately and incomplete combustion or corrosion from acidic by-products.

A recent development is the preparation of compost from mixtures of digested sludge and domestic refuse. The problems associated with this process are the resistance to biological breakdown of plastic wrappings, and the production of a sludge of the correct moisture level to give an economical operation. The advantages are the elimination of unsightly rubbish tips and the production of a saleable compost.

iii) Nitrogen Removal from Wastewaters

It is important to remove the nitrogenous constituents from the effluent from sewage treatment to minimize the toxicity of the water and to reduce its oxygen demand. The natural nitrogen cycle is utilized for this purpose, a simplified version of the cycle is given in Fig. 3.

The removal of nitrogenous material commences almost as soon as the sewage is formed. During passage through the sewer almost all the urea is decomposed to NH_3 and CO_2. The organic compounds of nitrogen (as micro-organisms or organic material) can be removed by sedimentation or precipitation. The inorganic nitrogen compounds can be removed, in the organic form, by bacterial or algal transformation. However, inorganic nitrogen can also be removed by nitrification and denitrification.

The presence of ammonia in an effluent is undesirable, because at a concentration of 10 mg/l at pH 8.5, rainbow trout are killed in about 8 h. Ammonia in a waste is first nitrified to either nitrite or nitrate by aerobic biological processes and then denitrified, in the absence of dissolved oxygen, to molecular nitrogen.

Nitrification is achieved in two stages by different species of autotrophic nitrifying bacteria:-

$$NH_4^{\oplus} + O_2 \xrightarrow[\text{species}]{\text{Nitrosomonas}} NO_2^{\ominus} + H^{\oplus} + H_2O$$

$$H^{\oplus} + NO_2^{\ominus} \longrightarrow HNO_2$$

and
$$NO_2^{\ominus} + O_2 \xrightarrow{\text{Nitrobacter} \atop \text{species}} NO_3^{\ominus}$$

The nitrifying bacteria will metabolize over a wide range of pH values, but the maximum rates for Nitrosomonas and Nitrobacter occur in the pH range 7-9 [56]. The extent of nitrification is usually less in winter than in summer, suggesting that the nitrifiers may be sensitive to temperature. Nitrification is inhibited by the presence of ammonia and nitrous acid [57] and by Cr, Zn, Cd and Pb [58].

iv) <u>Nitrification in the Activated Sludge Process</u>

If a significant population of nitrifying bacteria is present in the activated sludge, nitrification will take place during the aerobic treatment of organic wastes. There is a minimum aeration time, t_m, which must be exceeded to maintain a nitrifying culture [59]. Also the dissolved oxygen concentration must be 0.5-0.7 mg/l for nitrification to take place and will cease if it falls to 0.2 mg/l.

In any single passage through an activated sludge plant the increase in nitrifying bacteria must equal that of the heterotrophic population. If this does not occur then the nitrifying organisms are progressively lost from the system. The long aeration period, in the activated sludge process, ensures that the slow growing nitrifying bacteria are not replaced by the organisms which grow rapidly and metabolize the organic material.

Nitrification can also be controlled by the solids retention [60]. This is a measure of the average retention time of the cells in the system. If the solids retention time is maintained below the maximum generation time of the organisms, bacteria will be washed out of the system and the process fails. The process can be made reliable if nitrification is carried out in a separate aeration tank, which follows the one used for removing organic materials [61]. In this system, both stages can be operated under optimum conditions.

One of the first signs of an overloaded plant is the ceasing of nitrification. Nitrate production is also easily inhibited by heavy metals and this makes industrial sewage difficult to nitrify. The process is also inhibited by phenol and cresols, but not by ammonium thiocyanate (500 mg/l) [62].

v)   Nitrification and the Biological Filter

Nitrification takes place during the treatment of organic wastes by the biological filtration process. The biological filter can be a better nitrifying agent than the activated sludge plant because a greater segregation of species is provided than is the case in the general purpose activated sludge floc. When organic oxidation and nitrification occur in the same filter, the organic oxidation is normally seen at the top of the filter and nitrification at the bottom. Where dual filter systems are employed the organic material is oxidized in the first filter and nitrification occurs in the second. The effiency of the filter may be increased by increasing the depth of the filter, but this is not controllable nor capable of a high efficiency.

Nitrification can be carried out by a submerged filter system, and the design of these is discussed by Young and McCarty [63].

If complete removal of nitrogen is required, then the most promising way is to use biological denitrification of the effluents which have been nitrified. This approach becomes more important as it is thought desirable to remove all the nitrogen from wastewater to protect lakes, reservoirs and rivers from excessive plant growth.

Denitrification is a process by which $NO_3^{\ominus}$ and $NO_2^{\ominus}$ are reduced to molecular nitrogen. During the process nitrous oxide may be formed, but it does not appear as a significant intermediate during denitrification in aqueous solutions.

The ability to carry out denitrification is widespread and the bacterial genera Pseudomonas, Achromobacter and Bacillus are able to carry out the process. When molecular oxygen is absent they can use nitrates or nitrites as terminal electron acceptors when oxidizing organic material for energy. Certain autotrophic bacteria will carry out denitrification whilst oxidizing an inorganic energy source.

The amount and nature of the organic material may be a problem, because there is little control on these in municipal wastewaters [64], and industrial and agricultural wastewaters may not contain suitable electron donors. These factors can lead to inefficiency and unreliability. In order to provide reliability methanol has been used as a relatively cheap electron donor [65]

$$NO_3^{\ominus} + CH_3OH \longrightarrow NO_2^{\ominus} + CO_2 + H_2O$$

$$NO_2^{\ominus} + CH_3OH \longrightarrow N_2 + CO_2 + H_2O + OH^{\ominus}$$

The denitrification of agricultural wastewater has been carried out in California using anaerobic ponds and an anaerobic filter [66]. Also, anaerobic activated sludge has been used for the denitrification of municipal waste treatment plant effluents [67]. A series of four tanks in series: anaerobic stirred, aerobic stirred, anaerobic stirred and aerobic plug flow, have been used to remove both nitrogen and phosphate from effluent [68].

Disposal of Sewage from Coastal Towns

The discharge of sewage from coastal towns may take place directly into the sea without any treatment other than screening or comminution. If the outfall is close to the low-water mark revolting conditions may result if recognisable sewage solids accumulate on the beach, or are floating in the sea, or unpleasant odours are produced. Medical risks to bathers may be present and consumers of contaminated sea food may be at risk. There may also be adverse ecological effects or loss of amenity.

The health risks from sea bathing around the coasts of England and Wales can virtually be disregarded (M.R.C.). A few cases of para-typhoid fever associated with two grossly fouled beaches could have been due to contact of the patients with gross faecal solids [69].

The fish and shellfish taken from an environment polluted with sewage can be contaminated with pathogenic bacteria. The risk of infection to the consumer is eliminated by the minimum of heat treatment required in cooking. The health risk arises when the shellfish or fish are eaten raw, or only lightly cooked. If the sewage also contains industrial waste then fisheries and shellfisheries could be affected adversely by petrochemicals, phenolic compounds or heavy metals.

When sewage is discharged into the open sea from long outfalls, the overall biological productivity of the water is raised and the diversity and amount of bottom fauna is changed in the vicinity of the outfall. There are no widespread effects. However, if the opportunity for interchange with unpolluted water is limited severe effects can arise. There will be a substantial reduction in the diversity of bottom dwelling fauna and flora, destruction of shellfish beds and nuisance from the growth of algae.

The dispersion of sewage in the sea can be studied using sewage constituents e.g. coliform bacteria or phosphate, or by adding tracers e.g. ^{82}Br or easily identifiable bacteria such as Serratia indica or spores of

Bacillus subtilis var. niger. After release from the outfall the sewage
rises immediately to the surface because of its low density relative to
that of sea water. This forms the characteristic boil close to the outfall
at a position which remains relatively constant during slack water, but
wanders randomly when the tide is running strongly. Near the boil the
sewage is principally a layer at the surface about 1 m deep. The sewage
streams out from the boil in the direction of the tidal current and expands
in width fairly rapidly within 100-200 m from the outfall. It mixes with
the water below, and about 1,000 m from the outfall will be almost uniformly
distributed with respect to depth. Under constant conditions the number of
coliform bacteria in mixtures of sewage and sea water decreases approximately logarithmically with time. The death rate is increased significantly by increased sunlight, and to a lesser extent by increases in
salinity or temperature. The viable numbers fall by 90% in less than an
hour in samples at the surface in daylight, but such a reduction will take
several hours at night [70]. One of the main factors in deciding whether
to provide a full biological treatment before discharge is that the cost
and provision of full treatment on land can lead to loss of a land amenity.
If a count of 10^4 coliforms per 100 ml is required, this can be achieved by
discharging disintegrated sewage through an outfall 200 m in length, and
this is cheaper than giving a full treatment and discharging through a 50 m
outfall. As the standard of purity required is increased then the
difference between the alternatives diminishes, but even at 10^2 per 100 ml
it is still cheaper to discharge untreated sewage.

Septic Tanks and Cesspools

Small communities without mains sewerage systems often use septic
tanks or cesspools for the disposal of domestic sewage. If these are
properly constructed, well maintained and not overloaded they are very
effective.

A cesspool is a storage tank for sewage and it should be watertight.
It should not be allowed to overflow and when full it should be emptied for
the safe disposal of its contents elsewhere. If this system is used for a
household of four persons, and each person uses 30 gal of water a day, it is
a very expensive method of sewage disposal.

A septic tank provides a primary stage of sewage treatment and is
intended to overflow. It is a tank through which sewage is passed to
remove suspended solids, and in which the separated material is retained as

scum or sludge until it can be pumped out. It is often thought that a
septic tank gives complete treatment of the sewage i.e. that any solid
matter is completely digested or liquified if left in the tank long enough.
This is not so, much of the organic material needs further treatment by
aerobic biological digestion before it can be discharged into a watercourse.
Under optimum conditions about half of the material will be digested, so
unless the settled material is removed at intervals the tank fills and
there is no further settlement.

Polishing of Effluents from Sewage Works

The effluent, when it has passed through the sewage treatment, will
pass the Royal Commission standard and have a BOD less than 20 p.p.m. and
a suspended solids content less than 20 p.p.m. These levels may no longer
be adequate, and the effluent may need a polishing or tertiary treatment.
The main purpose of this treatment is to remove the suspended material
which has passed through the humus tank without sedimentation. Any suspended material usually accounts for most of the five day BOD of the
effluent. Bacteria have to be removed if the effluent is to be discharged
near the beds used for the cultivation of edible shellfish.

Rapid gravity sand filters provide a mechanical means of polishing an
effluent. These are similar to those used in drinking water purification
and consist of a bed of sand supported on an under-drainage system. They
are provided with a system for back washing and air scouring. About
72-90% of the suspended material in humus tank effluent can be removed in
this process [71] and the coli-aerogenes count reduced by 32% [72].

The slow sand filter is only suitable for small sewage works. These
consist of a layer of sand which rests on a layer of coarser material and
this rests on a system of drainage pipes. This type of filter has no back
washing facility. The effluent is allowed to percolate through the filter,
and when the head loss becomes excessive the surface layer of sludge has to
be removed manually. Up to 68% of the suspended solids and about 60% of
the coli-aerogenes organisms can be removed.

The micro-strainer was developed for use at waterworks but can be used
to treat humus tank effluent. This consists of a drum revolving on a horizontal axis, and it has the curved surface covered with a stainless steel
fabric of special weave. The whole revolves, at a speed up to 100 ft/min,
and the humus tank effluent enters the drum at one end and passes out
through the fabric; the other end being closed. This system will remove

up to 58% of the suspended solids [72].

The humus tank effluent can be run off on to grass plots, and this removes up to 73% of the suspended solids and 97% of the coli-aerogenes organisms. Alternatively, the humus tank effluent can be run into a lagoon. This method is very useful where the soil conditions are such that the grass mat is easily detached. The purification is brought about by a mixture of sedimentation and biological purification. The reduction in numbers of E. coli in lagoons is correlated with the level of biologically active u.v. light [73].

Activated charcoal has been suggested for polishing effluents, and the water discharged after this treatment can be used to supplement surface waters or run directly into a potable water reservoir [74]. It has been proposed that treatment of settled sewage with polyelectrolytes followed by activated charcoal treatment could replace the conventional biological processes.

Productive Use of Organic Waste

When resources were scarce during the Second World War there was considerable interest in using industrial waste waters productively. This interest waned, but has revived recently. This is particularly true of the paper industry which has become increasingly aware of its pollution problems. The spent sulphite liquor from the wood-pulping industry can be used for the microbial production of fumaric acid, ethanol, acetone, butanol, propanol, lactic acid, propionic acid, methane and hydrogen sulphide or the production of biomass e.g. yeasts [75]. The separation of the volatile organic acids can present problems, but the market price probably justifies this. Solid organic waste e.g. bagasse from sugar cane production can be hydrolysed and fermented to give ethanol which in turn can be used as a fuel in admixture with petrol.

Organic wastes can often be extracted to give protein which can be used in animal fodder e.g. waste from a potato starch plant can be precipitated to give 70-80% of crude protein [76] as can wastes from animal and vegetable processing plants [77]. Reverse osmosis has been suggested for the treatment of whey to provide a useful source of nutrients and remove a serious pollution problem [78].

Wastes which are rich in carbohydrate can be converted to single-cell protein by aerobic fermentation [79,80]. If the waste is low in phosphate or nitrogen then care must be exercised in adding these or a fresh pollution problem may be created. The choice of organism to produce single-cell protein may present problems, because bacteria will grow rapidly but their value as a food is not established, whilst algae require regions where there is adequate sunlight. The yeast Candida utilis is favoured because it is easily separated by centrifugation after fermentation. If filamentous fungi are used they are easy to separate from the liquor by filtration. After separation the cell mass must be heated to destroy viability without loss of the properties of the proteins. Inadequate drying may also cause problems of spoilage with undesirable micro-organisms.

Organic wastes from the food and drink industries have been investigated for their potential to produce single-cell protein. When waste from maize and pea processing plant was metabolized by the Fungi Imperfecti, up to 96% of the BOD was removed, in addition to all the ammoniacal nitrogen and inorganic phosphate. The yield (dry weight) was about 50% of the feed BOD, and the protein content of the mycelium was about 50% (dry weight) [81]. Similarly, about 80% of the BOD was removed from potato processing waste by a symbiotic culture of Endomycopsis fibuliger and Candida utilis [82]. The first organism produces an active amylase which converts starch into glucose and maltose, and the second grows actively and produces the cell mass. The yeast Candida ingens has been grown on the supernatant liquid derived from the anaerobic fermentation of monogastric animal waste [83].

Chemical wastes have been used as substrates for micro-organisms, for example, the waste from the production of artificial fibres will support the growth of algae [84], or the yeast Rhodotorula glutinis will grow on the waste from the pyrolysis of coal [85]. With starting materials of this nature there is always a possibility of the carry-over of toxic compounds.

If domestic sewage is used as a substrate for single-cell protein production, the cost of removing the organic material is greater than the use of conventional sewage treatments. Although the product can be sold, the recovery, from such a dilute nutrient source, is not high. The growth of algae on oxidation ponds is, of course, possible if sufficient sunlight is available, but is not a feasible proposition in the United Kingdom. Large quantities of cell protein are produced in the conventional sewage treatment processes as sludge, particularly as activated sludge. This material is nutritionally suitable for use as an animal foodstuff supplement [86], but it is still only used for methane production or as a fertilizer.

REFERENCES

1   Preliminary Assessment of Suspected Carcinogens in Drinking Water.
    U.S. Environmental Protection Agency. June 1975.
2   I.G. Wallis, Int. J. Environ. Stud., 6(1974)107.
3   J. Cairns, in B.A. Whitton (Editor), River Ecology, Blackwell, Oxford,
    1975, p. 588.
4   H.A. Hawkes, The Ecology of Waste Water Treatment, Pergamon Press,
    Oxford, 1963.
5   R.L. Bolton and L. Klein, Sewage Treatment: Basic Principles and
    Trends, Butterworths, London, 1961.
6   P. Coackley, Process. Biochem., 4(1969)27.
7   J.W. Abson and K.H. Todhunter, in N. Blakeborough (Editor), Biochemical
    and Biological Engineering Science, Vol. 1, Academic Press, London,
    1967, p. 309.
8   D. Herbert, Continuous Culture of Micro-organisms, Monograph No. 12,
    Society of Chemical Industry, London, 1961, p. 21.
9   W.O. Pipes, Adv. Appl. Microbiol., 8(1966)77.
10 E.B. Pike, E.G. Carrington and P.A. Ashburner, J. Appl. Bact., 35
    (1972)309.
11 B. Lighthart and R.T. Oglesby, J. Water Poll. Control Fed., 41(1969)
    R267.
12 L.A. Allen, J. Hyg., 43(1944)424.
13 F.F. Dias and J.V. Bhat, Appl. Microbiol., 13(1965)257.
14 J.E. Loveless and H.A. Painter, J. Gen. Microbiol., 52(1968)1.
15 R.F. Unz and J.A. Davis, J. Water Poll. Control Fed., 47(1975)185.
16 V.E. Leonova and V.F. Karpuktin, Microbiology, 43(1974)116.
17 C.R. Curds and A. Cockburn, Water Res., 4(1970)237.
18 T.H. Fairbrother and A. Renshaw, J. Soc. Chem. Ind. Lond., 41(1922)134.
19 C.R. Curds, A. Cockburn and J.M. Vandyke, Water Pollut. Control,
    67(1968)213.
20 J.M. Watson, Nature, 155(1945)171.
21 C.R. Curds, J. Gen. Microbiol., 33(1963)357.
22 W.B. Cooke and W.O. Pipes, Proc. 23rd Int. Waste Conf. Purdue Univ.,
    Engng. Extn. Ser. No. 132, 1969, p. 170.
23 U.L. Diener, G. Morgan-Jones, W.M. Hagler and N.D. Davis,
    Mycopathologia, 58(1976)115.
24 W.O. Pipes, Proc. 20th Int. Waste Conf. Purdue Univ., Engng. Extn. Ser.
    No. 118, 1965, p. 647.

25 T. Schofield, Water Pollut. Control, 70(1971)32.

26 P.H. Jones, Proc. 20th Int. Waste Conf. Purdue Univ., Engng. Extn. Ser. No. 118, 1965, p. 297.

27 R.E. Anderson, Water Pollut. Abstr., 42(1968)263.

28 A. Sladka and V. Zahradka, Morphology of Activated Sludge; Water Research Institute, Technical Paper No. 126, Prague-Podbaba, 1970.

29 M.W. Tenney and W. Stumm, J. Water Poll. Control Fed., 37(1965)1370.

30 G. Peter and K. Wuhrmann, Proc. 5th Int. Conf. Water Poll. Res., San Francisco, 1970. Pergamon Press, Oxford, 1971.

31 K. Crabtree, E.McCoy, W.C. Boyle and G.A. Rohlich, Appl. Microbiol., 13(1965)218.

32 H.A. Painter and A.P. Hopwood, Rep. Prog. Appl. Chem., 54(1969)405.

33 F.F. Dias and J.V. Bhat, Appl. Microbiol., 12(1964)412.

34 Y. Tezuka, Appl. Microbiol., 15(1967)1256.

35 Y. Tezuka, Appl. Microbiol., 17(1969)222.

36 W.H.J. Hattingh, Wat. Waste Treat., 9(1963)476.

37 F.F. Dias, N.C. Dondero and M.S. Finstein, Appl. Microbiol., 16(1968) 1191.

38 E.J.C. Curtis, Water Res., 3(1969)289.

39 C.R. Curds and A. Cockburn, J. Gen. Microbiol., 66(1971)95.

40 C.R. Curds and G.J. Fey, Water Res., 3(1969)853.

41 B.G. Oliver and E.G. Cosgrove, Water Res., 8(1974)869.

42 I. Radhakrishnan and A.K. Sina Ray, J. Water Poll. Control Fed., 46(1974)2393.

43 L. Vaicum and A. Eminovici, Water Res., 8(1974)1007.

44 A.J. Woodward, D.A. Stafford and A.G. Callely, J. Appl. Bact., 37(1974)277.

45 G. Fair, J. Geyer and D. Okum, Water and Wastewater Engineering, Vol. 2, John Wiley and Sons, New York, 1968, p. 34.

46 J.E. Germain, Water Pollut. Control Fed., 38(1966)192.

47 R.L. Antonie, Fixed Biological Surfaces - Wastewater Treatment, CRC Press, Cleveland, 1976.

48 W.J. Oswald, Proc. 6th Int. Conf. Water Pollut. Res., Jerusalem, Israel, 1972.

49 A.M. Buswell and H.F. Mueller, Ind. Engng. Chem., 44(1952)550.

50 W. Parker, J. Proc. Inst. Sew. Purif., (2)(1945)58.

51 R.L. Ward and C.S. Ashley, Appl. Environ. Microbiol., 31(1976)921.

52 F.M. Wellings, A.L. Lewis, and C.W. Mountain, Appl. Environ. Microbiol., 31(1976)354.

53 R.C. Sidle, J.E. Hook and L.T. Kardos, J. Environ. Qual., 5(1976)97.

54 Ministry of Agriculture, Fisheries and Food, Survey of Mercury in Food, H.M.S.O., London, 1971.

55 R.E. Brown, J. Water Poll. Control Fed., 47(1975)2863.

56 B. Boon and H. Laudelout, Biochem. J., 85(1962)440.

57 A.C. Anthonisen, R.C. Lochr, T.B.S. Prakasam and E.G. Srinath, J. Water Poll. Control Fed., 48(1976)835.

58 T.E. Wilson and M.D.R. Riddell, Water Wastes Eng., 11(1974)56.

59 A.L. Downing, T.G. Tomlinson and G.A. Trusdale, J. Proc. Inst. Sewage Purif., 3(1964)537.

60 A.W. Lawrence and P.L. McCarty, J. Am. Soc. Civil Engrs. San. Engng. Div., SA3(1970)757.

61 E.F. Barth, R.C. Brenner and R.F. Lewis, J. Water Poll. Control Fed., 40(1968)2040.

62 D.A. Stafford, J. Appl. Bact., 37(1974)75.

63 J.C. Young and P.L. McCarty, J. Water Poll. Control Fed., 41(1969)R160.

64 W.K. Johnson and G.B. Vania, Sanit. Engng. Rep. No. 175S, University of Minnesota, 1971.

65 P.L. McCarty, L. Beck and P. St. Amant, Proc. 24th Purdue Ind. Waste Conf. Lafayette, Ind., 1969, p. 1261.

66 T.A. Tamblyn and B.R. Sword, Proc. 24th Purdue Ind. Waste Conf. Lafayette, Ind., 1969, p. 1135.

67 K. Wuhrmann, in E.F. Gloyana and W.W. Eckenfelder (Editors), Advances in Water Quality Improvement, University of Texas Press, Austin, 1968.

68 J.L. Barnard, Water Wastes Eng., 14(1974)33.

69 Public Health Laboratory Service, J. Hyg., 57(1959)435.

70 A.H.L. Gameson and J.R. Saxon, Water Res., 1(1967)279.

71 A.E.J. Pettet, W.F. Collett and T.H. Summers, J. Inst. Sewage Purif., (1949)399.

72 Ministry of Technology, Notes on Water Pollution No. 22, H.M.S.O., London, 1963, p. 2.

73 J. Calkins, J.D. Buckles and J.R. Moeller, Photochem. Photobiol., 24(1976)49.

74 G.J. Stander and J.W. Funke, Chem. Eng. Prog., Symp. Series No. 78, 63(1967)1.

75 J.C. Mueller and C.C. Walden, Process Biochem., 5(1970)35.

76 Protein Recovery from Potato Starch, Process Biochem., 3(1968)51.

77 A.P. Hopwood and G.D. Rosen, Process Biochem., 7(1972)15.

78 I.K. Nielsen, A.G. Bundgaard, O.J. Olsen and R.F. Madsen, Process
    Biochem., 7(1972)17.

79 T.K. Ghose, Process Biochem., 4(1969)43.

80 A.J. McLoughlin, Process Biochem., 7(1972)27.

81 B.D. Church and H.A. Nash, U.S. Environmental Protection Agency,
    Wat. Pollut. Control Res. Ser. 12060 EDZ 08/71. U.S. Government
    Printing Office, Washington D.C., 1971.

82 M. Tveit, in M. Brook (Editor), Biology and the Manufacturing
    Industries, Academic Press, London, 1967, p. 3.

83 D.P. Henry, R.H. Thomson, D.J. Sizemore and J.A. O'Leary, Appl.
    Environ. Microbiol., 31(1976)813.

84 H.M. Palamar-Mordvyntseva, V.K. Marinich, V.V. Grabooskaya and
    S.M. Newhaus, Ukrainski Botanichnyi Zhurnal., 23(1966)56.

85 F. Leibnitz and R. Schulze, Continuous Culture of Micro-organisms,
    Proc. 2nd Symp. Prague, 1962; Czech Academy of Science, Prague, 1964.

86 S.C. Pillai, E.G. Srinath, M.L. Mathur, P.M.N. Naidu and
    P.G. Muthanna, Wat. Waste Treat., 11(1967)316.

DISINFECTION AND RECYCLING OF WATER

Disinfection of Water

The aim of all processes of disinfection of water is to destroy all potentially pathogenic micro-organisms so that the water, or waste, cannot transmit disease. If disinfectants are used in treating water, then any residual material should be easily detected so that quantitative measurements can be made on their levels. This is easy to fulfil if halogens are used, but other agents are more difficult, or impossible, to monitor. To cover these cases an estimate can be made using an indicator organism, e.g. *Escherichia coli*. This may not give a true assessment of the microbiological safety of water, because of the resistance of viruses to many disinfection processes.

Any disinfectant used for the treatment of drinking water must be nontoxic to humans. The most widely used drinking water disinfectants are chlorine or chlorine containing compounds. Chlorine and its compounds are strong oxidizing agents, and their disinfecting power can be easily removed in reactions with organic and inorganic compounds: when chlorine reacts with ammonia and organic compounds which contain nitrogen, chloramines are produced which retain some of the disinfecting powers. The reactivity of chlorine decreases as the pH increases and the reaction rate increases with rise in temperature. The addition of chlorine to water results in the formation of hypochlorous acid and hydrochloric acid:-

$$Cl_2 + H_2O \quad \rightleftharpoons \quad HOCl + H^{\oplus} + Cl^{\ominus}$$

This reaction is complete within seconds at pH values about 4 and at ordinary temperatures. There is little free chlorine in dilute solutions. Disinfection activity is associated with the hypochlorous acid which is formed and which dissociates:-

$$HOCl \quad \rightleftharpoons \quad H^{\oplus} + OCl^{\ominus}$$

The degree of dissociation is very dependent on pH, and temperature is less important. At pH values from 6 to 8.5 there is almost complete dissociation, but below 6 there is little dissociation. Therefore, to obtain good disinfection of water with chlorine the control of pH is very

important, because chlorine has little killing power above pH 8.5. In
potable waters the pH is such that chlorine exists both as hypochlorous
acid and hypochlorite ion, and these together are taken as free available
chlorine. The chloramines constitute combined available chlorine.
Residual chlorine in a wastewater may be toxic to aquatic animals [1].

The chlorination of wastewater may not eliminate all the potentially
pathogenic bacteria because lactose non-fermenters can survive the chlorination process [2]. However, Salmonella sp. are more sensitive to
0.3-1.0 mg/l of chlorine than E. coli [3]. Viruses on the other hand are
more resistant to chlorination than the bacteria used as indicator
organisms. The chlorination of wastewater can give rise to undesirable
compounds e.g. chlorinated biphenyl [4] which may be a pollution problem
downstream from the discharge point.

The use of chlorine dosage cartridges has been effective in treating
well water in Transcarpathia, and a level of 0.3-0.5 mg/l of chlorine is
maintained in the water. This treatment has been effective in reducing the
incidence of typhoid [5].

For the disinfection of small volumes of water, bleaching powder and
sodium hypochlorite are effective, or the compounds chloramine T, dichlor-
amine T and halazone may be used.

| Chloramine T | Dichloramine T | Halazone |

The solutions of these compounds are quite stable, but like chlorine they
have a reduced efficiency in the presence of organic material. Again the
action is due to the production of hypochlorous acid.

Chlorine dioxide is also an effective water disinfectant [6] and its
activity increases with rising pH. It is also less affected than chlorine
by the presence of organic materials. The cost has kept its usage down but
it is useful to control taste and odour [7].

The other halogens bromine and iodine have not been used for the disinfection of drinking water supplies on a large scale. The use of iodine for the disinfection of small scale supplies has been investigated.

Ozone has been used, particularly in France, as a water disinfectant [8] or for treating sewage effluent [9]. Because it is a strong oxidizing agent it has to be generated at the point of use, and there is little residual effect. Ozone has the advantage of removing tastes from water, including that of chlorophenol, and it virtually eliminates phenol at a concentration of 4-6 mol of ozone/mol of phenol [10]. Ozone does, however, impart a powerful penetrating taste of its own which may be more objectionable than that of chlorinated water. Ozone is a highly effective disinfectant which kills bacteria, viruses and cysts more rapidly than chlorine.

Other methods have been used for the disinfection of small scale water supplies including silver [11] which can be impregnated on filter elements or sand (Catadyn process). Ultra-violet light has been used for treatment of small water supplies, but the main limitation is that of poor pene- tration. The use of ionising radiations would overcome this.

The excess lime method is effective for disinfecting and purifying and, where necessary, softening water. The bactericidal level is about 1 to 2 parts calcium oxide/100,000 parts of water, the excess lime being removed by carbon dioxide.

Recycling of Water

To provide a measure of control over the safety of drinking water, the water supplies were regrouped in England and Wales by the Water Act of 1945 which ensured that each water supplying authority was large enough to employ qualified personnel. In the U.S.A. legislation was introduced, in the Safe Drinking Water Act of 1974, to improve the quality of water supplies.

In the United Kingdom, the control has extended to the planning stage by the establishment of the Water Resources Board in 1963 to cater for the future supplies of England as a whole. This function was taken over by the Central Water Planning Unit, under the Water Act 1973. This Act created nine regional water authorities for England and a national authority for Wales, with a single authority exercising control over an entire river basin. Their legal powers include extensive powers to lay down and enforce standards of quality for the effluents discharged into rivers. The creation of the water authorities should help in providing water of the optimum quality for drinking purposes.

Water supply authorities have usually attempted to find adequate supplies of unpolluted water, requiring the minimum of treatment, for domestic and industrial purposes. When a country becomes industrialized and heavily populated, it becomes more difficult to find such supplies and it is increasingly important to use water which is polluted. Water reclamation (or renovation) is practised on a large scale at the moment, when water supplies are taken from rivers which receive significant amounts of purified sewage effluent upstream from the abstraction point. The proportion of sewage effluent in a river may be very high during a drought period. The river Thames will contain about 14% sewage effluent under average weather conditions, and the river Lea a higher proportion. The only surface water available for Rotterdam is the river Rhine which receives over 70% of the total sewage of West Germany. A more extreme example is seen in Agra (India) which in times of drought has a water supply which is nearly all partially treated sewage from New Delhi, 190 km away [12].

In certain circumstances it may be essential to treat sewage effluents and use them immediately for industrial purposes. Separate water supplies can be provided for domestic and industrial use with the industrial supply being drawn from a river, and this may be of poorer quality than a sewage effluent. The provision of a separate network of mains (to provide drinking water or industrial water) in urban areas would be expensive, and if accidental cross-connections were made there would be a risk to health.

However, the use of sewage effluent to make good any losses in cooling towers is already a practical possibility [13]. When used for this purpose very little treatment is required, except to prevent algal growth (by chlorination), corrosion and foaming. The suspended solids content of the effluent must also be low. Similarly, sewage effluent can be used in a steel works [14].

The replenishing of natural waters occurs when treated sewage effluent is returned to a river. Modern water treatment makes it possible to purify most river waters to standards laid down by the World Health Organization [15,16] for drinking water, but the cost may be high. The best results are obtained when the treatment and control of the effluents entering a river is combined with adequate treatment of the water abstracted for domestic use. Where an international river, like the Rhine, is involved, there may be difficulties when pollutants are discharged into the river by a country not wishing to re-use the water.

Highly purified sewage effluent could be used without dilution to supply recreational lakes, and where treated wastewater is discharged into lakes it should be introduced below the trophogenic layer [17]. Percolation lagoons can be used to replenish the groundwater system. Also, where the rainfall is inadequate the use of sewage effluents for crop irrigation can be essential. Stainbridge [13] reported that in South Africa and South West Africa, of the 100 million gallons/day of effluent produced by local authorities, about one-quarter was used for industrial purposes and nearly half for crop irrigation. The problems which can arise from this are associated with the accumulation of certain elements in the soil or with the public health aspects. Crops irrigated in this way might have to be those which are cooked before consumption. The effect of detergent contamination on crops irrigated with wastewater has been investigated and a concentration of 1,600 mg/l of a heavy duty non-enzyme detergent stimulated the growth of maize [18]. This was probably due to the high phosphate level. At higher levels e.g. 8,000 mg/l, and enzyme containing detergents at 14,000 mg/l abnormal growth characteristics were observed. This damage was probably due to the level of sodium in the detergents. Detergents do not pose a problem in causing groundwater contamination [19] nor are the residues found in fish regarded as unsafe for human consumption [20].

The possibility of ecosystem irrigation as a means of groundwater recharge and improving water quality has been examined in respect of the river Trent, England. The Trent and its tributaries receive most of the domestic and industrial effluent of the Midlands, and in 1965 from a total flow of 1,600 million gallons/day at Nottingham about 264 million gallons/day were sewage effluent. Of this 40% was trade effluent discharged into the river via foul sewers, with a further 17% of trade effluent discharged directly into the river. The feasibility of producing potable water from the river at Nottingham has been examined by recharging the riverside gravels, or by constructing a river purification lake on the river Tame. A promising approach is the recharging of the Bunter sandstone by spray irrigation of natural heathland. Polluted river water can be sprayed onto the heathland and the water recharges the aquifer, with an improvement in water quality without any significant change in the balance of the plant species [21].

The provision of safe drinking water has lead to a decline in the deaths from waterborne pathogenic organisms. The effectiveness of the measures which have been adopted can be seen by the fact that eleven cities on the Ohio watershed were still drawing water in 1954 without purification from the same source as in 1906. The average death rate from typhoid fever was 76.8/100,000 in 1906 and was still 74.5/100,000 in 1945. Another sixteen cities had since 1906 started to treat their water supplies without any change in the source of supply. In these cities the death rate from typhoid fell from 90.5 to 15.3/100,000. The procedures are well established for sand filtration and were used in London in 1829, and the use of chlorine was introduced in 1910. The combination of filtration and chlorination provides the basic treatment for the purification of river-derived drinking water. The treatment frequently includes preliminary coagulation or flocculation, followed by sedimentation to remove most of the suspended solids, and this may be followed by storage of the raw water as an additional safeguard. These methods with a control of residual chlorine or ozone levels has virtually eliminated bacterial waterborne disease in the United Kingdom. There may be some virus survival, although there is no evidence for this.

Enteric viruses are present in wastewater and effluent all the year round, and polio I and echo 4 are prevalent [22]. Viruses will survive discharge into seawater [23,24] and the presence of particulate matter influences the survival [25]. The enteric bacteria can survive the pressures of deep-sea conditions (500 to 1,000 atmospheres). However, there is no evidence of accumulation of terrestrial organisms on the sea bed [26]. The usual indicator organisms do not die off so there could be serious long term ecological and health problems. The detection of E. coli in seawater is complicated because the presence of salt inhibits its growth [27].

The problems of providing safe drinking water are now concerned with trace amounts of chemicals or other substances. A problem which has only recently been recognized is that antibiotic resistant coliform organisms which appear in the water environment could become the agents by which pathogenic bacteria acquire the genetic characteristics which confer drug resistance. This genetic information is transferred by plasmids [28], and plasmid transfer of antibiotic resistance (R factors) can occur in some coliforms (isolated from a sewage sludge bed) during conjugation with a recipient strain of Salmonella gallinarum [29]. Hospital sewage contains more R-factor coliform organisms and many more of these organisms had multiple R-factors than coliforms isolated from domestic sewage, or other sources [30]. However, it was calculated that less than 5% of the R-factors in sewage originate from hospitals. A potentially more serious situation is the high frequency (73%) of enterobacteria resistant to the antibiotic Ampicillin which were detected in the River Stour, Kent, during 1974. In the same river in 1970 only 27% of the organisms were Ampicillin resistant [31]. The appearance of these antibiotic resistant organisms and the transfer of this resistance to pathogenic organisms may have a serious effect on the long term usefulness of antibiotic therapy.

The significance of the long term ingestion of pollutants in water is difficult to determine. Their effects may not be immediately apparent and can be screened by other factors which may have similar effects. In this context it may be significant that much of London takes its drinking water from the rivers Thames and Lea which are contaminated with wastewater. However, about 15% of the population of southeast London have an unpolluted water supply from the Kent chalks. The population which obtains its water from the Kent chalks has a lower cancer mortality than the population served by river waters [32]. Also, studies carried out in Holland showed that where towns take drinking water from polluted rivers there is a higher death rate from cancer than where underground sources are used [33].

The problem is complicated because new chemical compounds are being introduced into the environment and their long term effects are unknown. The conventional water treatment is not totally effective in removing them and they may cause cancer, genetic damage or congenitally deformed children [34]. For example, of the 60 compounds identified in drinking water in the U.S.A., 2 were classed as extremely toxic, 16 very toxic, 14 moderately toxic, 1 non-toxic and the effects of the remainder unknown [35]. If oils or phenols are implicated in water pollution, these are readily detected by taste and the water is rendered unpalatable. Surface waters may contain small amounts of polycyclic aromatic hydrocarbons which are carcinogenic. Concentrations of these in treated drinking water rarely exceeds 0.2 μg/l, and this level has been adopted by the World Health Organization as an upper limit, because this gives a small intake when compared with that from other sources.

Where activated carbon is used as a treatment then many of the potentially dangerous chemicals are removed [36].

The efficient use of water resources, including wastewater, requires careful planning on more than a parochial scale, and in the United Kingdom there will have to be a considerable extension of the use of river-derived water for drinking purposes. There will also have to be close monitoring of the water, in toxicology and microbiology as well as chemical analysis.

148

REFERENCES

1   W.A. Brungs, J. Water Poll. Control Fed., 45(1973)2180.

2   E.N. Davis and S.R. Keen, Hlth. Lab. Sci., 11(1974)268.

3   M.S Belakovskii, Gig. Sanit., 4(1976)96.

4   R. Johnson, in J.L. Buckley (Editor), Natl. Conf. on Polychlorinated
    Biphenyls, Washington Environ. Protect. Agency, 1976, p. 379.

5   R.I. Grishchenko, Gig. Sanit., 10(1975)92.

6   G.M. Ridenour, R.S. Ingols and E.H. Armbruster, Water Sew. Works,
    96(1949)279.

7   R.N. Aston, J. Am. Water Wks. Assn., 42(1950)151.

8   D.C. O'Donovan, J. Am. Water Wks. Assn., 57(1965)1167.

9   S. Miller, B. Burkhardt, R. Ehrlich and R.J. Petersen, in Ozone
    Chemistry and Technology, Adv. Chem. Ser. Vol. 21, Amer. Chem. Soc.,
    Washington D.C., 1959, p. 381.

10 J.P. Gould and W.J. Weber, Water Pollut. Control Fed., 48(1976)47.

11 R.L. Woodward, J. Am. Water Wks. Assn., 55(1963)881.

12 World Health Organization, Tech. Rep. Ser. W.H.Ö. No. 517, 1973.

13 H.H. Stainbridge, J. Proc. Inst. Sew. Purif., 1965, p. 20.

14 P.H. McGauhey, in P.C.G. Isaac (Editor), Waste Treatment, Pergamon
    Press, London, 1960, p. 429.

15 World Health Organization, European Standards for Drinking Water,
    2nd ed., W.H.O., Geneva, 1970.

16 World Health Organization, International Standards for Drinking Water,
    3rd ed., W.H.O., Geneva, 1971.

17 H. Buchner and H. Ambucht, Schweiz Z. Hydrol., 37(1975)347.

18 Bull. Water Resources Res. Center Virginia Polytech. Inst., 62(1973)50.

19 W.F. Kaufman, J. Am. Water Wks. Assn., 66(1974)152.

20 S. Bellasai and S. Sciacca, Ig. Mod., 66(1973)384.

21 M.J. Chadwick, K.J. Edworthy, D. Rush and P.J. Williams, J. Appl. Ecol.,
    11(1974)231.

22 N. Buras, Water Res., 10(1976)295.

23 S. De Flora, G.P. de Renzi and G. Badolati, Appl. Microbiol., 30(1975)
    472.

24 S.A. Berry and B.G. Norton, Water Res., 10(1976)323.

25 C.P. Gerba and G.E. Schaiberger, Water Pollut. Control Fed., 47(1975)93.

26 J.A. Baross, F.J. Hanus and R.Y. Morita, Appl. Microbiol., 30(1975)309.

27 J.A. Papadakis, J. Appl. Bact., 39(1975)295.

28 R.C. Clowes, Sci. Amer., 228(1973)19.

29 L.K. Koditschek and P. Guyre, Marine Poll. Bull., 5(1974)71.

30 J. Linton, J. Med. Microbiol., 7(1974)91.

31 C. Hughes and G.C. Meynell, Lancet (ii)(1974)451.

32 P. Stocks, Regional and Local Differences in Cancer Death Rates, No. 1.
    London General Register Office, 1974.

33 S.W. Tromp, Schweiz Z. Path., 18(1955)929.

34 D.A. Okun, J. Am. Water Wks. Assn., 61(1969)215.

35 R.G. Tardiff and M. Deinzer, Toxicity of Organic Compounds in Drinking
    Water. 15th Water Quality Conf. Org. Matter in Water Supplies, Univ.
    Illinois, Urbana, 1973.

36 Water Research Association, Activated Carbon in Water Treatment, Water
    Research Association, Medmenham, Marlow, Bucks., 1974.

EUTROPHICATION

Before entering into a discussion of polluted ecosystems, it is necessary to define what is meant by a non-polluted system.

In a non-polluted community, there will be a large number of inter-dependent species. The relationship between any two species present may vary from pathogenic or predator, to symbiotic, to no direct relationship at all between the two species under consideration. In the non-polluted system, each species is represented by a relatively small and stable number of individuals.

Cairns and Lanza [1] define pollution as the appearance of some environmental factor for which the exposed community has inadequate information, and is therefore incapable of making an adequate and appropriate response.

One of the first effects noted in a polluted ecosystem is that the number of different species is progressively reduced, but the number of individuals per species is increased. In the final stages of pollution, it is not uncommon to find only a few species represented by a vast number of individuals.

The most important point is that as the range of species present in the ecosystem becomes more reduced, food chains become shorter and simpler and the community becomes more biologically unstable.

The question of population size raises one of the major problems of pollution studies, that is the quantification and reproducibility of data. The difficulties encountered have been discussed [2] and include problems such as obtaining a standard sample from a standard source, and the identification of some of the species involved. The term 'standard source' is considerably more complex than is frequently realized, as many micro-organisms will show variation in spacial distribution in response to factors such as light, turbidity, food supply, sex and/or age of the organism. The authors [2] report results in terms of an average number ($\pm$ 40%) for some of the most common species. As the more uncommon species are considered, then the reproducibility of the results will fall, as the total number of organisms under consideration will no longer be as statistically significant.

The growth of micro-organisms in any environment is controlled by a number of factors, the most important of which is the availability of the food and energy supply to the organism. Any physical or chemical factor changing the food or energy supply will cause a change in the microbial population. The type and magnitude of the change will determine whether or not the ecosystem becomes polluted.

The term 'eutrophication' is derived from a Greek word meaning 'rich', and eutrophic waters have been defined as those which have become enriched with organic or inorganic nutrients containing phosphate, and to a lesser extent nitrogen in some form. These nutrients allow the excessive growth of micro-algae whose metabolism causes the depletion of oxygen in water to produce the primary effect of eutrophication. Secondary effects include foul smells and tastes, caused by the generation of sulphides, extensive fish kills and the death of non-resistant micro-organisms. The final result is the total colonization of the water by a small number of resistant species and the build-up of sediment which, in the case of a lake, eventually causes it to become completely filled.

In a stable biological community there is a balance between the activity of producer organisms and consumer organisms. That is the relationship between photosynthesis (reduction of carbon dioxide to organic compounds by the use of solar energy) and the aerobic degradation of organic material is a steady state one. Odum [3] has produced a simplified stoichiometric equation to account for this.

$$106 \ CO_2 + 16 \ NO_3^- + HPO_4^= + 122 \ H_2O + 18 \ H^+ + energy$$

$$photosynthesis \ \circlearrowleft \ respiration$$

$$(C_{106}H_{263}O_{110}N_{16}P) + 138 \ O_2$$

algal protoplasm

In the Atlantic the ratios of dissolved nitrate:phosphate:carbonate are approximately 16:1:106 which reflects the values shown in the equation. As photosynthesis proceeds, dissolved phosphate and nitrate become exhausted more or less simultaneously, thus determining an upper limit to productivity in the sea. Some lakes have been shown to have similar simple stoichiometric relationships [4].

A steady state relationship between photosynthesis (P) and respiration (R) is necessary to maintain a constant chemical environment. The condition

THE PHOSPHORUS CYCLE

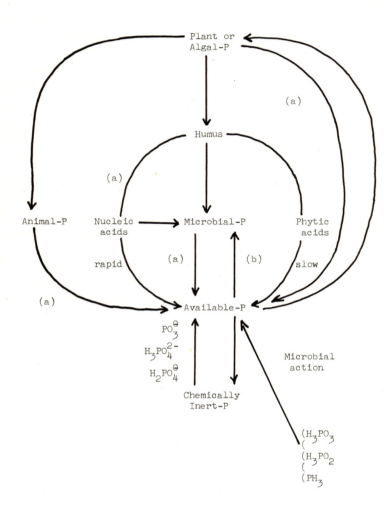

a = mineralisation
b = immobilisation

required for the synthesis of proteins and enzymes (biological catalysts).

Nitrogen and phosphate transformations can be split into three areas: (a) entry into the body of water, (b) metabolic reactions in the water and (c) removal from the system.

The origins of phosphate and nitrogen can be summarized as follows: (1) rain, rivers and streams, (2) agricultural and urban effluent and (3) regeneration of mineral deposits and turnover of the ecosystem. In addition to these three sources, nitrogen can also enter the system by means of nitrogen fixation and leave by the process of denitrification. These processes are known as the phosphate and nitrogen cycles. The first is considered here, whilst the nitrogen cycle, denitrification and nitrogen fixation are considered in more detail in the chapter on Sewage Treatment.

Much of the nutrient entering a lake will do so from agricultural or municipal wastes [19,20,21,22]. In many cases the waste will enter the water in an untreated form. Municipal wastes entering sewage treatment plants are subjected to secondary treatment which removes most of the organic matter, but only about 30% of the phosphate and 20% of the nitrogen. The rest of the nutrients enter the water supply and are an extensive cause of eutrophication. However, the amount of phosphate and nitrogen in a surface water cannot be determined solely by a knowledge of the amounts entering the lake due to the considerable exchange processes involved. The amount of dissolved oxygen can alter the dissolved phosphate considerably, reducing conditions causing the precipitation of phosphate as salts of calcium, iron or manganese [23,24] and the level of dissolved phosphate will thus be temperature dependent to some extent.

If untreated municipal or farm waste is allowed to enter the surface water, eutrophication is not the only problem. Many pathogenic (disease causing) micro-organisms are associated with human and animal sewage. These include enteric viruses such as those causing polio, and enteric bacteria causing diseases such as cholera and typhoid. In certain tropical and sub-tropical countries protozoans such as Entamoeba histolytica causing amoebic dysentery, or parasitic metazoans causing diseases such as schisto-somiasis (bilharzia) [25] may be found. These parasitic metazoans frequently possess complicated life cycles involving passage through one or more intermediate hosts such as snails or fish, and in many cases prevention or reduction of such diseases could result from improved engineering and drainage techniques.

The major organisms responsible for eutrophication are the micro-algae.

It should not be assumed that the photosynthetic ability of these organisms
will increase proportionally with the increase in phosphate and nitrogen.
Many other factors are involved such as temperature, light intensity,
presence of organic chemicals which may be used as an energy supply,
presence of toxic chemicals and the grazing activities of various micro-
herbivores or zooplankton. All these factors will be inter-related in a
complex manner and an alteration in one may cause widespread changes in the
community.

The micro-algae found in eutrophic situations can be divided into a
number of taxonomic groups, the most important of which are the
Cyanophyceae or blue-green algae.

The term 'blue-green algae' is a misnomer as this group is in fact a
specialized group of procaryotes containing photosynthetic pigments. The
type of photosynthesis is similar to that found in the true algae rather
than true bacterial photosynthesis [26], but on structural grounds the
Cyanophyceae must be considered to be procaryotes [27,28]. The structure
and identification of the Cyanophyceae are considered by Fritsch [29].

Although there are a large number of species of Cyanophyceae, the
literature on eutrophication is concerned with about twenty species capable
of forming populations sufficiently large to be called blooms. These
blooms last for varying periods of time. Barica [30] followed algal
growth in shallow self-contained lakes and showed that dense blooms
occurred which caused an oxygen deficiency in the water leading to a
collapse of the bloom. This would cause the release of nitrogen and phosphate nutrients and another bloom would occur [31].

The causes of bloom succession are subject to a certain amount of
controversy, and the subject is considered in a literature review by Welch
[32].

The species capable of forming water blooms are found in a number of
different groups in the Cyanophyceae, but one feature which most of them
possess is a gas vacuole. There are certain exceptions to this, amongst
them the genus Synechococcus, which is very narrow.

When studied in the electron microscope the gas vacuoles appear as
stacks of hollow membranous sub-units [33] which have been called gas
vesicles and which are stacked together to look like the cells of a honey-
comb. The vesicle walls are an insoluble protein which give a structural
rigidity capable of withstanding approximately two atmospheres pressure.
At pressures higher than this the vesicles collapse.

Measurements obtained in Anabaena flos-aquae show that the volume of the gas vacuoles increases as the culture ages, and results suggest that the role of the vesicles is to provide positive buoyancy enabling the algae to float. Dinsdale and Walsby [34] have shown that when A. flos-aquae is placed in a medium of high light intensity, the hydrostatic pressure of the cell rises and the gas vacuoles collapse allowing the cell to sink to a less intensely illuminated area. The critical factor deciding whether or not a bloom floats is thus its resistance to high light intensities. A secondary factor may however be the depletion of nutrients in the high population occurring in water blooms. One problem [35], unexplained as yet, is the fact that many cells removed from water blooms appear senile and are not capable of reproduction under laboratory conditions.

This flotation mechanism allows the cell to reach neutral buoyancy and explains the stratification of blue-green algae into different bands found in various lakes [36,37]. Some species produce gas vesicles in the reproductive cells, and in these cases the role is probably a dispersal mechanism allowing the cell to rise, float away from the parent and establish a new colony. The structure, function and physiology of the gas vacuoles of the Cyanophyceae is reviewed by Walsby [38].

The formation of gas vacuoles appears to be the only common structural feature amongst bloom forming blue-greens, although they do have several physiological features in common. Fitzgerald et al. [39] showed that 2,3DNQ (2,3-dichloro-1,4-naphthoquinone) is selectively toxic to bloom forming Cyanophyceae, and addition of this compound leads to bleaching and cell lysis [40]. They also appear to be more sensitive than other algae to copper and copper sulphate, which have been used effectively to control blooms of this type [41].

Several authors have suggested that there is a correlation between bloom formation, and the concentration of dissolved organic nitrogen the previous month, although the role of the organic material is not certain [31,35,42] as several authors have shown that organic nitrogen is taken up more slowly than inorganic nitrogen [43,44]. The organic compounds may act as a source of metabolites or vitamins for the algae enabling them to grow faster, or alternatively it may act by allowing the growth of mucilaginous bacteria thus raising the level of dissolved carbon dioxide [45] and favouring algal photosynthesis. This last factor may be of importance as Goldman et al. [10] have shown that the rate limiting factor of the growth of Scenedesmus is the level of dissolved carbon dioxide, although this

organism is not a member of the Cyanophyceae. The suggestion has also been
made that the organic material may act as a chelating or buffering agent.

Attempts to correlate the density of a bloom with the level of phosphate are complicated by the fact that many bloom forming species can
accumulate large reserves of polyphosphate which can be used to supplement
environmental phosphate [46]. Shapiro [47] has shown that phosphate uptake
is influenced by the type and concentration of anions present, and there
also appear to be considerable anomolies between field observations and
laboratory results regarding the optimum phosphate levels for some species.

Many of the Cyanophyceae are able to fix nitrogen, that is they can
convert atmospheric nitrogen to ammonia by means of the enzyme nitrogenase.
This means that theoretically nitrogen can never be the rate limiting
factor of growth of the blue-green algae. Nitrogenase has a wide substrate
specificity and is capable of reducing a number of compounds. The assay
normally used is the reduction of acetylene [48], although a recent paper
shows that this assay is not applicable under all conditions [49]. The
earliest stable end product is ammonia and species vary in the hydrogen
donor used, some using molecular hydrogen whilst others use organic compounds. In all cases nitrogen reduction is energy requiring.

The most common forms of nitrogen fixing Cyanophyceae are the filamentous forms containing large specialized cells known as heterocysts [50],
although some unicellular species also carry out the reaction [51]. There
has been a certain amount of controversy over the role of heterocysts, but
it is now accepted that they are the site of nitrogen fixation [52,53].

The observation has been made on several occasions that Cyanophyceae
produce chemicals toxic to other species [54,55], and the bacterial genus
Bacillus has been used to degrade geosmin, an odour producing compound
synthesized by Anabaena circinalis [56]. The maximum release of organic
matter has been shown to be during the log phase of growth and not after
the death of the cell. Tassignay and Lefevre [57] showed that water from
which blooms of Aphanizomenon gracile and Oscillatoria planctonica had
been removed was inhibitory to the growth of other algae.

The toxicity of these algae to other algae is not well documented but
members of the Cyanophyceae have been responsible for diseases in man and
domestic animals in several countries. Organisms implicated in the outbreaks are mainly species of Microcystis, Anabaena and Aphanizomenon.
Gastro-enteritis and various allergic conditions have been noted after
heavy blooms in water supplies. A number of toxins have been implicated

but only a few have been characterized, for example M. aeruginosa produces
a cyclic polypeptide with a molecular weight of approximately 2,600, and
A. flos-aquae produces an alkaloid of molecular weight 3,000.

Cyanophyceae are not the only group of bloom forming algae although
they are the most common.

Certain marine dinoflagellates may develop into blooms giving the sea
a red or brown colour. These blooms can lead to the death of fish and
other species. The toxin bearing dinoflagellates are consumed by shellfish
upon which they have no effect, but when the shellfish are eaten they cause
severe gastric complaints that may result in death. The dinoflagellates
causing these blooms are usually members of the genus Gonyaulax. The toxin
produced by G. catanella is saxotoxin with a molecular weight of approximately 370 and a lethal dose of about 9 $\mu g/kg$ of body weight. The organism
responsible for the 1968 bloom off the Northumbrian coast which caused the
deaths of large numbers of sea birds and sickness amongst a number of
people was G. tamarensis. Conditions leading to the formation of red
blooms are discussed by Pingree et al. [58].

The chrysophyte Prymnesium parvum which is found in brackish water
also produces a toxin, which has been known to cause the deaths of large
numbers of fish in Holland, Denmark and Israel.

The unicellular flagellate Crytomonas and the armoured dinoflagellates
Peridinium and Ceratium also cause blooms, and it has been suggested that
high turbidity favours the growth of flagellated algae over the
Cyanophyceae [59].

The type of algae found in a surface water may be used as an indicator
of the degree of pollution present. The major advantage of the algae over
organisms such as bacteria or zooplankton is that they are relatively easy
to count and identify [29]. Palmer [60] has compiled an extensive list of
algae species considered to be tolerant to pollution, and from these he
produced an algal pollution index which could be used to grade the degree
of organic pollution.

The control of eutrophication raises a number of possibilities. The
most obvious method is to control the amount of nitrogen and phosphates
entering the water system. The literature has been reviewed recently by
Foehrenbach [61], but it has to be accepted that in many cases it is both
practically and economically unrealistic to attempt to completely remove
these chemicals.

Larsen et al. have reported [62] that the removal of 70% of the phosphate entering Lake Shagawa, Minn. through tertiary treatment of sewage has caused a recovery to commence eighteen months after treatment started. The recovery of a lake after dredging has also been reported [63].

A number of initial experiments have been carried out [64,65] on growing micro-algae on wastes of various types, and in one case the algae were then used as foodstuff for livestock. A number of problems face this technique, one of which is harvesting the algae on a sufficiently large scale. Centrifugation is too complicated and micro-filters cause the build-up of slime. It was concluded that the best method is sand filters [66]. Other problems are drying the algae sufficiently to enable the resulting foodstuffs to be stored, and there is also the problem of toxic metabolites to be overcome. The use of secondary effluent mixed with sea water to grow macro-algae has been explored by Goldman et al. [67], whilst McShan et al. [68] used a combination of algae and brine shrimps.

The use of algicides to control algae has been reviewed by Sladeckova and Sladecek [69]. One very important point made by the authors is that if the effect of the algicide is too acute it may cause even greater problems. If all the algae are killed more or less simultaneously, then the decomposing cells may overload the capacity of the lake to oxidize them. The major problems associated with the use of algicides are the large volume of water involved, and possible long term effects of cumulative algicides on other species.

Peterson et al. [70] have considered the possibility of harvesting aquatic macrophytes in an attempt to control eutrophication. The wet weight of plants removed over the growing season was considerable, but the phosphorus input of the lake was only reduced by approximately 1.4%.

The use of natural pathogens of algae to exert biological control over their numbers has not been widely exploited, although several authors [71,72] have considered the possibility.

Fungal pathogens of the Cyanophyceae are mainly chytrids which appear to have a very limited host range, and in some cases even be confined to one specific structure in the host organism [73,74].

The effect of zooplankton grazing on the phytoplankton is considerable although most zooplankton prefer to feed on algae other than the Cyanophyceae, which may be due to the production of toxic metabolites by this group.

A number of authors have demonstrated the existance of viruses capable of lysing blue-green algae [75,76,77], and the subject is reviewed by Padan and Shilo [78]. Morphologically these virus are similar to the bacteriophages attacking bacteria and have been called cyanophages. The cyanophages appear to exhibit a considerable degree of host specificity, and it has been postulated that they may be responsible for the collapse of blooms of Cyanophyceae [79].

Bacterial pathogens of micro-algae are also known [80,81], the group mainly involved being the Myxobacterales. They tend to be much less host specific than the viruses, and will attack a wide range of host algae causing lysis of vegetative cells, but it is usually found that heterocysts and spore cells are resistant.

The use of natural pathogens in the biological control of algal blooms is as yet in its infancy. Work has however been carried out using viruses or bacteria [82] in field tests to control blooms. It is probable that the preferred pathogen will be a bacterial species as they cause lysis more quickly, will attack a much wider range of host cells, and the algae exhibit a lower probability of developing resistance.

REFERENCES

1   J. Cairns and G.R. Lanza, in R. Mitchell (Editor), Water Pollution
    Microbiology, Wiley Interscience, 1972, p. 245.
2   R.W. Edwards, B.D. Hughes and R.W. Read, in M. Chadwick and
    G.T. Goodman (Editors), The Ecology of Resource Degradation and
    Renewal, Blackwell, 1975, p. 139.
3   P.E. Odum, in T.A. Olson and F.J. Burgess (Editors), Pollution and
    Marine Ecology, Wiley Interscience, 1967.
4   W. Stumm and E. Stumm-Zollinger, in R. Mitchell (Editor), Water
    Pollution Microbiology, Wiley Interscience, 1972, p. 11.
5   P.E. Odum, Science, 164(1969)262.
6   R.G. Wetzel, Limnology, W.B. Saunders Company, 1975.
7   L.N. Vanderhoef, C.Y. Huang, R. Musil and J. Williams, Limnol. and
    Oceanog., 19(1974)119.
8   W.E. Miller, T.E. Maloney and J.C. Greene, Water Res., 8(1974)667.
9   M. Halmann and M. Stiller, Limnol. and Oceanog., 19(1974)774.
10 J.C. Goldman, W.J. Oswald and D. Jenkins, J. Water Poll. Control Fed.,
    46(1974)554.

11 L.L. Harms, J.N. Dornbush and J.R. Andersen, J. Water Poll. Control
    Fed., 46(1974)2460.

12 R.A. Vollenweider, Organization for Economic Co-operation and
    Development, Paris, 1968.

13 E. Gorham, J.W.G. Lund, J.E. Sanger and W.E. Dean, Limnol. and
    Oceanog., 19(1974)601.

14 J.R. Jones and R.W. Bachmann, Verh. Intl. Ver. Limnol., 19(1975)904.

15 T. Ahl, ibid. p. 1125.

16 C. Serruya, ibid. p. 1357.

17 W.D.P. Stewart, S.B. Tuckwell and E. May, in M.J. Chadwick and
    G.T. Goodman (Editors), The Ecology of Resource Degradation and
    Renewal, Blackwell, 1975, p. 57.

18 J.R. Jones and R.W. Bachmann, J. Water Poll. Control Fed., 48(1976)2176.

19 L.L. Harms, J.N. Dornbush and J.R. Andersen, J. Water Poll. Control
    Fed., 46(1974)2460.

20 J.W. Kluesener and G.F. Lee, J. Water Poll. Control Fed., 46(1974)920.

21 L. Stone, J. Amer. Water Works Assn., 66(1974)489.

22 G.L. Dugan and P.H. McGauhey, J. Water Poll. Control Fed., 46(1974)2261.

23 C. Serruya, M. Edelstein, U. Pollingher and S. Serruya, Limnol. and
    Oceanog., 19(1974)489.

24 W.A. Norvell, Soil Sci. Soc. Amer. Proc., 38(1974)441.

25 M. Muller, New Scientist, 69(1976)18.

26 G.E. Fogg, W.D.P. Stewart, P. Fay and A.E. Walsby, The Blue-green
    Algae, Academic Press, 1973, Ch. 8, p. 143.

27 R.Y. Stanier, M. Doudoroff and E.A. Adelberg, General Microbiology,
    McMillan, 3rd ed., 1971.

28 T.D. Brock, in N.G. Carr and B.A. Whitton (Editors), The Biology of
    Blue-green Algae, Blackwell, 1973, Ch. 24, p. 487.

29 F.E. Fritsch, The Structure and Reproduction of Algae, Vol. II,
    Cambridge University Press, 1945, p. 768.

30 J. Barica, Arch. Hydrobiol., 73(1974)334.

31 B.D. Vance, J. Phycol., 1(1965)81.

32 E.B. Welch, J. Water Poll. Control Fed., 48(1976)1335.

33 C.C. Bowen and T.E. Jensen, Science, 147(1965)1460.

34 M.T. Dinsdale and A.E. Walsby, J. Exp. Bot., 23(1972)561.

35 G.E. Fogg, Proc. Roy. Soc. B, 173(1969)175.

36 U. Zimmermann, Schweig Z. Hydrol., 31(1969)1.

37 G.E. Fogg and A.E. Walsby, Mitt. Internat. Verein Limnol., 19(1971)182.

38 A.E. Walsby, in N.G. Carr and B.A. Whitton (Editors), The Biology of Blue-green Algae, Botanical Monographs, Vol. 9, Blackwells, Oxford, 1973, Ch. 16, p. 340.

39 G.P. Fitzgerald, G.C. Gerloff and F. Skoog, Sewage and Industrial Waste, 24(1952)888.

40 B.A. Whitton and K. MacArthur, Arch. Mikrobiol., 57(1973)147.

41 G.P. Fitzgerald, Eutrophication Information Programme Literature Review, No. 2, Water Resources Centre, University of Washington, U.S.A., 1971.

42 C.E. Boyd, Arch. Hydrobiol., 73(1974)361.

43 P.A. Wheeler, B.B. North and G.C. Stevens, Limnol. and Oceanog., 19(1974)249.

44 D.M. Schell, Limnol. and Oceanog., 19(1974)260.

45 W. Lange, Nature (London), 215(1967)1277.

46 W.D.P. Stewart and G. Alexander, Freshwater Biol., 1(1971)389.

47 J. Shapiro, Water Res., 2(1968)21.

48 W.D.P. Stewart, in D.F. Jackson (Editor), Algae, Man and the Environment, Syracuse University Press, 1968, p. 53.

49 J.A.M. de Bont and E.G. Mulder, Appl. and Environ. Microbiol., 31(1976)640.

50 W.D.P. Stewart, Phycologia, 9(1970)261.

51 R. Rippka, A. Nielson, R. Kunisawa and G. Cohen-Bazire, Arch. Mikrobiol. 76(1971)341.

52 P. Fay, W.D.P. Stewart, A.E. Walsby and G.E. Fogg, Nature (London), 220(1968)810.

53 S. Bradley and N.G. Carr, J. Gen. Microbiol., 96(1976)175.

54 C.K. Lin, Hydrobiologia, 39(1972)321.

55 L.A. Mel'nichenko, I.G. Goronovskii, A.P. Potemskaya and I.A. Sakevich, Gidrobiol. Zh., 9(1973)82.

56 L.V. Narayan and W.J. Nunez, J. Amer. Water Works Ass., 66(1974)532.

57 M. Tassigny and M. Lefevre, Mitt Int. Verein Theor. Angew. Limnol., 19(1971)26.

58 R.D. Pingree, P.R. Pugh, P.M. Halligan and G.R. Forster, Nature, 258(1975)672.

59 A.J. Horne and A.B. Viner, Nature, 232(1971)417.

60 C.M. Palmer, J. Phycol., 5(1969)78.

61 J. Foehrenbach, J. Water Poll. Control Fed., 47(1975)1538.

62 D.P. Larsen, K.W. Malueg, D.W. Schults and R.M. Brice, Kogai To Taisaku, 11(1975)397.

63 L. Bengtsson, S. Fleischer, G. Lindmark and W. Ripel, Verh. Intl. Ver. Limnol., 19(1975)1080.

64 W.J. Oswald and C.G. Golutke, in D.F. Jackson (Editor), Algae, Man and the Environment, Syracuse University Press, 1968, p. 371.

65 D.C. Vanderpost and D.J. Torrien, Water Res. (G.B.), 8(1974)593.

66 E.J. Middlebrooks, D.B. Porcella, R.A. Gearheart, G.R. Marshall, J.H. Reynolds and W.J. Grenney, J. Water Poll. Control Fed., 46(1974)2676.

67 J.C. Goldman, K.R. Terone, J.H. Ryther and N. Corwin, Water Res., 8(1974)45.

68 M. McShan, N.M. Trieff and D. Grajcer, J. Water Poll. Control Fed., 46(1974)1742.

69 A. Sladeckova and V. Sladecek, in D.F. Jackson (Editor), Algae, Man and the Environment, Syracuse University Press, 1968, p. 441.

70 S.A. Peterson, W.L. Smith and K.W. Malueg, J. Water Poll. Control Fed., 46(1974)697.

71 R.S. Safferman, in D.F. Jackson (Editor), Algae, Man and the Environment, Syracuse University Press, 1968, p. 429.

72 R.S. Safferman, in N.G. Carr and D.A. Whitton (Editors), The Biology of the Blue-green Algae, Blackwell, 1973, Ch. 11, p. 214.

73 H.M. Canter, Trans. Brit. Mycol. Soc., 37(1954)111.

74 H.M. Canter, Trans. Brit. Mycol. Soc., 46(1963)208.

75 R.A. Levin, Nature (London), 186(1960)901.

76 R.S. Safferman and M.E. Morris, Science, 140(1963)679.

77 M.J. Daft, J. Begg and W.D.P. Stewart, New Phytol., 69(1970)1029.

78 E. Padan and M. Shilo, Bact. Revs., 37(1973)343.

79 R.S. Safferman and M.E. Morris, J. Amer. Water Works Ass., 56(1964)1217.

80 M. Shilo, Science J., (Sept.), 2(1966)33.

81 M.J. Daft and W.D.P. Stewart, New Phytol., 70(1971)819.

82 G.E. Fogg, W.D.P. Stewart, P. Fay and A.E. Walsby, The Blue-green Algae, Academic Press, 1973, p. 296.

THERMAL   POLLUTION

The use of power in the industrialized world is increasing very
rapidly.   The various power plants available depend to a large extent upon
the use of surface waters as a coolant, and these are then returned to the
environment at an elevated temperature.   The problem of thermal pollution
will be exacerbated to a considerable extent by the increasing use of
nuclear power stations which require approximately 50% more cooling water
than non-nuclear power stations of the same capacity.   It has been estimated that if present trends continue in the U.S.A., by 1980 the power
industry will be using more than 20% of the available fresh water supply.

Changes in temperature of a surface water are one of the simplest
parameters to measure, as it does not require sophisticated or expensive
apparatus.   Temperature changes cannot however be measured in isolation to
other parameters, as temperature fluctuations will cause both chemical and
physical changes in the water.

Variations in temperature will cause alterations in pH due to changes
in ionization and increased solubility or precipitation of bottom deposits.
This may have a considerable effect upon the organisms present as Brock
[1] has reported that bacteria grow well in boiling hot springs at neutral
or alkaline pH, whereas in acid hot springs at pH 2 the upper limit is
$70^{o}C$.   Similarly, algae grow at approximately $75^{o}C$ in neutral and alkaline
springs, but will only grow up to $55^{o}C$ in acid hot springs.

There will also be considerable changes in the oxygen content of
water over a wide temperature range.   For example, the oxygen content of
water at $90^{o}C$ is only 2% of the value of the oxygen content at $20^{o}C$.

The density of water is maximum at $4^{o}C$ and temperature alterations
from this value will cause a fall in the density and viscosity of the
water.   This will have a direct effect upon the flotation mechanisms and
movements of algae, zooplankton and fish.   This effect will probably be
greatest in the larval stages and when dispersal is taking place.

A fact frequently not considered is that the effect of thermal
pollution will alter depending upon the time of the year.   The temperature
of water which is not thermally polluted will vary depending upon the
length of the day.   The number of daylight hours will in turn affect the
amount of photosynthesis taking place, and thus the amount of oxygen being
produced.   If thermal pollution takes place in winter, the shorter daylight

hours will reduce the amount of plant photosynthesis taking place and oxygen production will fall. The increase in temperature will however cause increased respiration and an increased use of oxygen leading to anaerobic conditions being produced more rapidly. Thermal pollution leading to a given temperature will thus have a greater effect, in terms of an increased rate of eutrophication, in winter than in summer.

Biochemically, the result of increasing the temperature is an increase in the catalytic rate of enzyme activity, which approximately doubles for a rise in temperature of $10^{\circ}C$ between the approximate range of $0-40^{\circ}C$.

The effect of thermal pollution is thus two-fold, an increased catalytic rate and a reduced oxygen tension, giving a considerable increase in the rate of eutrophication over a long term.

The effect of thermal stress on micro-organisms has been reviewed by Farrell and Rose [2] and the upper temperature limit varies depending upon the type of micro-organism involved. Brock [1] gives an upper temperature limit for protozoa, animals and fungi of approximately $50^{\circ}C$, whereas non-photosynthetic heterotrophic bacteria have been isolated from boiling pools. Photosynthetic procaryotes such as the blue-green algae have an upper temperature limit of approximately $75^{\circ}C$. The ability of an organism to tolerate high temperatures decreases as its molecular organization increases, and elevated temperatures thus favour procaryotes. It is extremely important, however, to distinguish between active growth at an elevated temperature and survival at the same temperature. The question of the evolution of thermophily has been reviewed by Mitchell [3].

Most naturally occurring thermal areas such as the hot springs and geysers of Yellowstone only contain a small number of species [4]. The feeding chains are greatly reduced and simplified, thus producing a highly unstable biological environment, which is extremely susceptible to minor fluctuations. In the thermal areas analysed by Brock [4] the standing crop of organisms was highest at about $55^{\circ}-60^{\circ}C$. At higher temperatures the crop is reduced due to thermal limitations, whereas below this temperature the grazing activities of herbivores becomes significant. The growth of the Cyanophyceae in thermal environments has been considered by Castenholz [5], and Castenholz and Wickstrom [6] have reviewed the subject of thermal streams.

Small rises in temperature may be beneficial to a community if the ambient temperature is low. Cherry et al. [7] showed that a rise of $3-4^{\circ}C$ above the ambient produced an increase in total bacterial count and also an increased diversity of species. A rise of $10^{\circ}C$ however caused an increased bacterial count, but a reduction in the diversity of species.

Although a small increase in temperature at the lower end of the range may be beneficial, the same increase at the upper end of the tolerance range will cause a reduction in the number of species. This reduction will be found especially in the vertebrate and invertebrate fauna. In the algae there will be a shift from diatom dominated communities to those dominated by Cyanophyceae [8], although a review by Anraku [9] considers that the progressive change is from diatoms to green algae to blue-green algae.

Small temperature increases in the region $70-75^{\circ}C$ will cause a shift from photosynthetic procaryotes to heterotrophic procaryotes. The large effect of a small temperature rise in the upper range is caused by the fact that for many tropical and sub-tropical species the lethal temperature is only a few degrees above the optimum temperature for growth [4].

Data from research on power station effluents has shown that if fish are adequately protected, then so are the invertebrate and procaryotic flora and fauna. Coutant and Talmage [10] report that a minimum of two years ecological and biological survey is necessary before siting a new power plant. A temperature of approximately $40^{\circ}C$ would appear to be the maximum tolerated by fish species.

Bader et al. [11] have shown that a sustained temperature above $33^{\circ}C$ causes a high mortality rate of macro-algae and sea grass leading to a major loss of food and increased erosion. Algae are generally considered to be more resistant to temperature changes than vascular plants [12], although temperature has been shown to have considerable effects on algal metabolism, photosynthesis, reproduction and distribution [13,14,15]. Coutant and Talmage have reviewed the subjects [10]. Thermal adaptation in yeast species has been considered recently [16].

The problems of thermal pollution may not be direct. Two recent papers consider the effect of temperature on chemical toxicity towards aquatic species and generally conclude that there is a great lack of information [17,18].

The literature is somewhat contradictory, but many authors consider that bacterial and viral infections of fish are greater at high temperatures. Fish which have spent winter in a thermal area are in poorer condition than those kept at the normal ambient temperature, even though the fish in the thermal area were feeding whereas the others were not [19]. Thermal effluents in reservoirs have been shown to cause increased

several authors have proposed model analytical systems to predict changes due to thermal pollution, or plant shut-down.

All these aspects of thermal pollution have been considered in the following reviews [10,22,28].

REFERENCES

1   T.D. Brock, Science, 158(1967)1012.

2   J. Farrell and A.H. Rose, in A.H. Rose (Editor), Thermobiology, Academic Press, 1967, Ch. 6, p. 147.

3   R. Mitchell, Quat. Revs. Biol., 49(1974)229.

4   T.D. Brock, Ann. Revs. Ecol. Systems, 1(1970)191.

5   R.W. Castenholz, in N.G. Carr and B.A. Whitton (Editors), The Biology of Blue-green Algae, Blackwell, 1973, Ch. 19, p. 379.

6   R.W. Castenholz and C.E. Wickstrom, in B.A. Whitton (Editor), River Ecology, Blackwell, 1975, Ch. 12, p. 264.

7   D.S. Cherry, R.K. Guthrie and R.S. Harvey, Water Res., 8(1974)149.

8   M. Hickman, Hydrobiologia, 45(1974)199.

9   M. Anraku, Bull. Plankton Soc. Jap., 21(1974)1.

10 C.C. Coutant and S.L. Talmage, J. Water Poll. Control Fed., 47(1975) 1656.

11 R.G. Bader, M.A. Roessler and A. Thorhaug, in M. Ruivo (Editor), Marine Pollution and Sea Life, Food and Agriculture Organization, Fishing News (Books) Limited, London, 1972.

12 B.A. Whitton, J. Zool. Soc., 29(1972)3.

13 F.E. Round, in D.F. Jackson (Editor), Algae, Man and the Environment, Syracuse University Press, 1968, Ch. 6, p. 73.

14 T.D. Brock, Nature, 214(1967)882.

15 M. Hickman and D.M. Klarer, Brit. Phycol. J., 10(1975)81.

16 H. Arthur and K. Watson, J. Bact., 128(1976)56.

17 J. Cairns, A.G. Heath and B.C. Parker, J. Water Poll. Control Fed., 47(1975)267.

18 P.D. Abel, J. Fish Biol., 6(1974)279.

19 R.R. Massengill, Chesapeake Sci., 14(1973)138.

20 R.T. Sawyer and D.L. Hammond, Biol. Bull., 145(1973)373.

21 R.R. Avtalion, Z. Malik, E. Lefler and E. Katz, Bamidgeh, 22(1970)33.

22 C.C. Coutant and H.A. Pfuderer, J. Water Poll. Control Fed., 46(1974) 1476.

23 J.M. McKim, R.C. Anderson, D.A. Benoit, R.L. Spehar and G.N. Stokes, J. Water Poll. Control Fed., 48(1976)1544.

24 G.R. Ash, N.R. Chymko and D.N. Gallup, J. Fish Res. Board Can., 31(1974)1822.

25 D.W. Blinn, J. Phycol., 9(1973)4.

26 B.G. Joyner and T.E. Freeman, Phytopathology, 63(1973)681.

27 R. Singleton and R.E. Amelunxen, Bact. Revs., 37(1973)320.

28 C.C. Coutant and S.S. Talmage, J. Water Poll. Control Fed., 48(1976) 1486.

THE SULPHUR CYCLE AND WASTE RECOVERY

Sulphur is an essential element in the metabolism of all living organisms, and most organisms consume or transform only relatively small quantities of compounds such as sulphate. This is used in the synthesis of the amino acids cysteine and methionine, which are found in proteins.

Sulphur Cycle

There are two types of bacteria which biochemically transform amounts of sulphur that are much in excess of the requirements for biosynthesis of cellular sulphur compounds. These organisms are the dissimulatory sulphate-reducing bacteria and the chemolithotrophic sulphur-oxidizing bacteria. In the natural environment they occur close together and can produce a dynamic state of alternate oxidation and reduction of sulphur compounds in soil or water (Fig. 1).

Fig. 1.

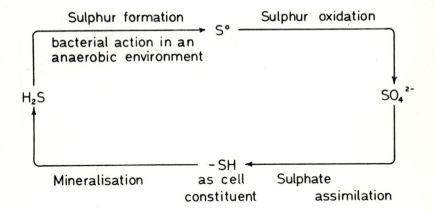

Sulphate-reducing Bacteria

These organisms belong to the genera _Desulfovibrio_ and _Desulfotomaculum_ and grow anaerobically, using sulphate instead of oxygen for respiration The metabolism of the sulphate-reducing bacteria has been reviewed by Postgate [1].

Sulphate-reducing bacteria are responsible for the deposition of several sulphide ores, including iron (as hydrotroilite or pyrites), copper, lead and possibly others. The formation of pyritic fossils is typical of the process which accompanies their action. In many cases the role of sulphate-reducing bacteria in metal sulphide deposition has to be checked by isotope fractionation. The isotope ratio used is that of $^{32}S:^{34}S$ which ranges in terrestrial sulphides between 21.3-23.2:1. Sulphur in meteors has a ratio of 22.2:1 and this is used as a standard for other ratios. When Desulfovibrio reduces sulphate it forms sulphide preferentially from ^{32}S and uses ^{34}S more slowly. The deviation produced in the sulphide level is easily measured by mass spectrometry. The sedimentary sulphides which are possibly of biological origin are similarly enriched, whilst igneous sulphides show less deviation from the meteoric standard.

A magnetic iron sulphide has been reported to be formed by Desulfovibrio [2] and metal sulphide precipitation may occur within the bacteria or on their surface. Precipitation of the sulphide, under experimental conditions, is more rapid and more complete when the bacteria are present than when bacteria-free mixtures of metal and sulphide ions are prepared.

The role of Desulfovibrio in the production of copper sulphide was questioned because the organism is very sensitive to copper ions. However, the copper in the soil is mainly in a non-ionic form and presumably non-toxic, and actively growing Desulfovibrio would precipitate the copper as sulphide and so detoxify it.

Sulphide can be produced by bacterial fermentation and a British project to do this was discontinued when it reached the stage of economic feasibility [3]. The process was attractive because it used sewage sludge as reducing agent. The work was at the pilot plant stage [4,5,6] and the details of yields, extraction, disposal and associated corrosion problems had been considered. The process will probably be revived when world resources of sulphur are sufficiently depleted. The advantage of a sulphate-reducing fermentation over a methane fermentation for treating sewage sludge is the decreased water content of the digested product. Also, the sulphate-reducing bacteria can be used to treat certain industrial wastes e.g. distillery and citric acid wastes [7,8].

The sulphate-reducing bacteria can also be the cause of certain problems because they are implicated in metal corrosion in aquatic environments [9]. The growth of Desulfovibrio in waterlogged pulp wood produces thiolignin and this will neutralize the mercurial fungicides which are added later in paper manufacture causing rotting to occur [10].

Sulphur Oxidizing Bacteria

Many bacteria can oxidize sulphur, thiosulphate and other reduced forms of inorganic sulphur to sulphate. However, two types are totally dependent upon such oxidations and can produce relatively large amounts of sulphate during their growth. These are the chemolithotrophic Thiobacillus species and the photosynthetic sulphur bacteria e.g. the green pigmented Chlorobium and the red pigmented Chromatium.

The photosynthetic species develop under anaerobic conditions when light is available to provide the source of energy for growth. During photosynthesis in muds and lake waters they may oxidize the sulphide generated by the sulphate-reducing bacteria and precipitate sulphur on such a scale that it can be economically mined. For example, in one of the Slavyansk Lakes, Ukr SSR, H_2S is formed most vigorously by the sulphate-reducing bacteria in the upper layer of the bottom sludge at a rate of 15.6 mg H_2S/kg damp sludge/day [11]. This H_2S is oxidized mainly by · Chlorobium phaeovibrioides.

The thiobacilli are more important in the turnover of sulphur and they occur in soils, muds, lakes and oceans. They use the oxidation of sulphur compounds as their energy source for growth, and they have to oxidize large amounts of them in order to multiply.

The work relating to the enzymatic basis for sulphur oxidation has been reviewed [12,13]. The exact mechanism for the oxidation of reduced sulphur compounds by these organisms has not been completely elucidated and different species of thiobacilli may have different pathways. The overall oxidation scheme is summarized below:

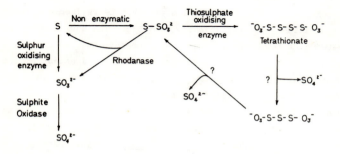

In soils where sulphur fertilizers have been added, the thiobacilli can increase soil acidity to such an extent that the range of vegetation supported is completely altered. This feature can be exploited if a more acid soil is required. The processes are the same as those responsible for acid mine drainage.

Acid Mine Drainage

If the sulphide ores are left in their natural state they do not affect the environment, but if they are exposed to the air they can undergo a series of changes, in which microbial attack plays an important part. These changes can give rise to a water pollution problem of considerable dimensions. The type of polluted water resulting from mine drainage depends on the type of mine and the nature of the geological surroundings.

The acid drainage from coal mines often has a pH value less than 4.0 and contains a significant amount of soluble sulphate and iron. The water from mines can have values less than pH 2.9 with sulphate concentrations ranging from 3,600 to 41,700 p.p.m.; the water with the highest sulphate level was pH 1.4 [14]. The extent of the pollution from coal mining can be extensive. The River Trent area of the United Kingdom includes a region which produces one-third of the U.K. coal production. This gives pollution of the river, which receives the discharge of mine water, coal washing plant effluents and the runoff from spoil tips. Any drainage from coal stocking sites can also be polluting. Some 100,000 m^3/day of mine water is discharged from seventy-six mines. These discharges can have chloride levels up to 36,000 mg/l, compared with the chloride level in the sea which is 19,000 mg/l [15]. Depending on the salinity of the mine water and the level in the receiving stream, values for salinity up to 1,200 mg/l have been recorded. This can have a profound effect on the flora of the stream, and the iron present may precipitate as ferric hydroxide and blank out weed growth and the macro fauna.

The levels of certain cations e.g. zinc, copper, cadmium, aluminium and arsenic, in mine drainage can reach concentrations which are toxic to aquatic life. The effects can spread many miles downstream and result in the death of fish [16]. Also micro-organisms can be affected and if the drainage is discharged into a sewage works, the biological processes may be inhibited [14]. The toxic levels reached in streams will bring about a large change in the protozoan, algal and bacterial communities, allowing species which are acid tolerant, or able to oxidize sulphur and iron to predominate. This problem does not finish when the mining operations stop, but will continue with leaching for a considerable time.

Acid mine waters contain a predominance of iron-oxidizing bacteria (Thiobacillus ferro-oxidans) which may reach 10^6 organisms/ml, and sulphur-oxidizing bacteria (Thiobacillus thio-oxidans). The autotrophs of alkaline waters are mainly non-acidophilic sulphur-oxidizing bacteria [17].

The problem is not confined to mining operations but is also found when land is reclaimed from the sea, because the underwater soil is exposed to oxygen. This results in an increase in soil acidity and the production of soluble iron from iron sulphide and soluble cations. This can be a major problem in parts of the Netherlands.

Sulphur mining can give similar problems, for there can be sufficient sulphur, from the dust, for the soil bacteria to oxidize the element and so give acidity. The soil pH can fall to less than 4 and plants will fail to develop [18].

The mechanism by which acid waters are formed from the metabolism of sulphide ores may involve three stages [19]. These are the oxidation of sulphide to sulphate:-

$$2FeS_2 + 7O_2 + 2H_2O \longrightarrow 2FeSO_4 + 2H_2SO_4$$

the conversion of ferrous iron to ferric:-

$$4FeSO_4 + 2H_2SO_4 + O_2 \longrightarrow 2Fe_2(SO_4)_3 + 2H_2O$$

and the precipitation of ferric hydroxide:-

$$Fe_2(SO_4)_3 + 6H_2O \longrightarrow 2Fe(OH)_3 + 3H_2SO_4$$

Overall the reaction products from FeS_2 and oxygen, would be $Fe(OH)_3$ and sulphuric acid.

Thiobacillus ferro-oxidans can oxidize insoluble sulphides and iron, but the formation of an acid mine water may not be a simple bacterial oxidation of FeS_2. The problem is one of deciding which reaction steps are chemical and which bacteriological. It has been shown that FeS_2 is oxidized slowly in the absence of micro-organisms, but the bacterial action contributes by oxidizing ferrous iron to ferric:-

$$4Fe^{2+} + O_2 + 4H^\oplus \longrightarrow 4Fe^{3+} + H_2O$$

This production of ferric ions can bring about the non-biological oxidation of FeS_2:-

$$14Fe^{3+} + FeS_2 + 8H_2O \longrightarrow 15Fe^{2+} + 2SO_4^{2-} + 16H^{\oplus}$$

which will account for the observed acidity. The rate limiting step in this process is ferrous oxidation [20]. This means that the bacterial oxidation could be critical at this stage, as the ferrous oxidation is very slow below pH 4.5. Walsh and Mitchell [21] stated that T. ferro-oxidans is very active only at pH values less than 3.5 and they suggested that Metallogenium, which is an acid tolerant, filamentous iron oxidizer, initiates the process in the range 3.5 to 4.5. This increases the acidity and Metallogenium dies out and T. ferro-oxidans, which is now near its optimum, takes over. Even if the water is neutralized it is usually unacceptable to the water authorities because of its high iron content.

Microbial Mining

Microbial attack on metal sulphides has been exploited in several countries, including the U.S.A., Canada and Australia, for the process of waste dump leaching. This process relies on the thiobacilli for the recovery of economically important metals, particularly copper [22].

$$\text{Cu}_2\text{S} + \tfrac{1}{2}\text{O}_2 + \text{H}_2\text{SO}_4 \xrightarrow{\text{T. ferro-oxidans}} \text{CuS} + \text{CuSO}_4 + \text{H}_2\text{O}$$

Then $$\text{CuS} + 2\text{O}_2 \longrightarrow \text{CuSO}_4$$

When minerals, such as copper sulphides, are mined, only high-grade ores, containing amounts of metal that can be removed economically are generally processed by the mechanical and chemical extraction procedures. For example, in Utah, the Kennecott Copper Corporation amasses about 250,000 tons of waste rock every day. This results in the building of slag heaps containing millions of tons of copper. This can be recovered by the simple process of pumping water onto the dumps and collecting the effluent liquor which is a dilute solution, with impurities, of copper sulphate (see above). The liquid is collected in storage dams of about 20 million gallons capacity. The copper is recovered by the addition of metallic iron (as scrap) and precipitation of the copper. This process gives about 5% of the total copper output of the major companies. This process can also be used to leach nickel and copper from copper-nickel deposits [23], and by leaching uranium from impoverished ore the yield can be increased by up to 30% [24]. Vanadium and molybdenum can be recovered in a similar fashion.

<u>Thiobacillus ferro-oxidans</u> can also be utilized in the renewal of uranium by the transformation:

$$U^{4+} + 2Fe^{3+} \longrightarrow U^{6+} + 2Fe^{2+}$$

All these processes provide a cheap, slow method of upgrading low-grade ores at acid pH values.

Unwanted impurities can be removed from minerals using <u>T. ferro-oxidans</u> leaching e.g. coking coal can be upgraded if its pyrites content is lowered.

Use of Power Station Ash

Pulverised fuel ash (PFA) from power stations can give good plant growth if suitably treated with a thin layer of soil. When it leaves the boiler PFA is sterile but it obtains a microbial population from the water used for lagooning, or conditioning, and from the atmosphere. Micro-organisms will grow in PFA and so they may be able to improve its ability to support plant growth. A problem arises because of the variation in mineral content of the ashes, the coal from South Wales, and so the ash, is low in boron but the Midlands coal is high in boron. The former is easily colonised by plants, but the latter is generally phytotoxic.

Rippon <u>et al</u>. [25] showed that when micro-organisms were added to PFA in pot experiments there was an improvement in plant growth. However, the value of bacterial fertilizers, in field trials, was not proved. The addition of organic waste, as poultry manure, domestic refuse or sewage, to PFA did improve the fertility. As PFA ages it is colonised by micro-organisms until the population resembles that of soil.

REFERENCES

1   J.R. Postgate, Bact. Rev., 29(1965)425.
2   A.M. Freke and D. Tate, J. Biochem. Microbiol. Technol. Eng., 3(1961)29.
3   K.R. Butlin, S.C. Selwyn and D.S. Wakerly, J. Appl. Bact., 23(1960)158.

4   S.C. Burgess and J.A. Scott, Ann. Rep. Sci. Advisor, 1957, London County Council, 1958, p. 7.

5   S.C. Burgess and J.A. Scott, Ann. Rep. Sci. Advisor, 1958, London County Council, 1959, p. 8.

6   S.C. Burgess and J.A. Scott, Ann. Rep. Sci. Advisor, 1959, London County Council, 1960, p. 9.

7   J. Barta, Continuous Cultivation Micro-organisms, Symp. Prague, 1962, p. 325.

8   J. Barta and E. Hudcova, Folia Microbiol. (Praha), 6(1961)104.

9   G.H. Booth, A.W. Cooper and A.K. Tiller, J. Appl. Chem., 13(1963)211.

10   P. Russell, Chem. Ind., (1961)642.

11   E.N. Chebotarev, V.M. Gorlenko and V.I. Kachalkin, Microbiology, 42(1973)475.

12   P.D. Kelly, Aust. J. Sci., 31(1968)165.

13   P.A. Trudinger, Rev. Pure Appl. Chem., 17(1967)1.

14   L.V. Carpenter and L.K. Hendon, W. Va. Univ. Eng. Exp. Sta. Res. Bull., 10(1933)

15   W.F. Lester, in B.A. Whitton (Editor), River Ecology, Blackwell, Oxford, 1975, p. 489.

16   W.R. Turner, Progr. Fish Cult., 20(1958)45.

17   R. Tabita, M. Kaplan and D. Lundgren, Third Symp. Coal Mine Drain., Bituminous Coal Res., Monroeville, Pa., 1970, p. 94.

18   M. Kvol, W. Malliszewska and J. Suita, Pol. J. Soil Sci., 5(1972)25.

19   R.A. Brant and E.Q. Moulton, Ohio State Univ. Eng. Exp. Sta. Bull., 1960, p. 179.

20   P.C. Singer and W. Stumm, Science, 167(1970)1121.

21   F. Walsh and R. Mitchell, Environ. Sci. Technol., 6(1972)809.

22   H. Sakaguchi, A.E. Torma and M. Silver, Appl. Environ. Microbiol., 31(1976)7.

23   S.A. Moshnyakova, G.I. Karavaiko and E.V. Shchetinina, Microbiology, 40(1971)959.

24   F.F. Barbich and B.V. Krainchanich, Microbiology, 41(1972)303.

25   J.E. Rippon and M.J. Wood, in M.J. Chadwick and G.T. Goodman (Editors), Ecology of Resource Degradation and Renewal, 15th Symp. Br. Ecolog. Soc. 1973, Blackwell, Oxford, 1975.

OIL POLLUTION

The subject of oil pollution is probably one of the most emotive subjects in any discussion on environmental pollution. It is not usually realized, however, that considerable amounts of oil and hydrocarbon material have always found their way into the ecosystem.

This may arise as seepage from oil fields following geological disturbances or it may be hydrocarbon material formed biosynthetically. Many plants have been shown to produce hydrocarbons from 16 to 33 carbon atoms long [1,2,3]. Kaneda [4] has shown that hydrocarbons in spinach exist in two distinct ranges, the group on the external surface of the leaf being odd-numbered with $\underline{n}\text{-}C_{31}$ predominating, and the group occurring internally having a significant proportion of even-numbered hydrocarbons ranging from $\underline{n}\text{-}C_{16}$ to $\underline{n}\text{-}C_{28}$.

Some algae [5] contain significant quantities of $\underline{n}$-pentadecane and $\underline{n}$-heptadecane and many bacteria possess hydrocarbon material [6,7] as do certain insects.

Branched chain hydrocarbons are produced biosynthetically, the main ones containing iso and ante-iso methyl groups [8]. Large amounts of pristane (2,6,10,14-tetramethyl-pentadecane) are also found in the ecosystem, presumably derived from the phytol group of chlorophyll [9].

In addition to these, many organisms produce unsaturated hydrocarbons of the terpene type which give rise to the C_{30} unsaturated hydrocarbon squalene, the precursor of the steroids. Also synthesized in considerable amounts are the cyclic hydrocarbons of the β-carotene type which function in plant photosynthesis.

The amount of hydrocarbon released into the sea is given by Hughes and McKenzie [10] and Floodgate [11] who consider that the marine biomass is responsible for the synthesis of a greater total amount of hydrocarbon than any other source.

The ecosystem has always been able to cope with naturally occurring amounts of hydrocarbon, mainly because it is spread over a wide area, diluted in extremely large volumes of sea water and usually released gradually over a period of time.

The major problem faced by the environment at the present is dealing with massive local pollution over a short time period following accidents such as that of the Torrey Canyon or the oil well leak at Santa Barbara.

The most recent of these major catastrophes is the sinking of the Argo Merchant off the Massachusetts coast in late 1976. This accident spilt $7\frac{1}{2}$ million gallons of oil into the sea which has contaminated some of the richest commercial fishing grounds off the north east coast of the U.S.A. The problems of drilling and oil spills on the outer Atlantic shelf have also been considered by Travers and Luney [12].

To date, accidents of this type have tended to happen in areas close to the coast and the major problem has been to stop the oil before it reached the coast line. Major oil spillages on land are rare and most problems of oil spillage have been associated with water pollution.

The position may change in the future with the construction of the trans-Alaskan oil pipe lines. Evidence available suggests that oil spilt on land in temperate or hot climates is degraded within six months. However in soils such as Alaska where the metabolic activity of soil micro-organisms is low, the persistence of oil can be measured in terms of years. In a recent paper, Raymond et al. [13] have shown that oil degradation in soil at a site in Oklahoma was negligible between November and March. The degradation of petroleum under Artic conditions has recently been con-sidered [14].

One major difference in pollution on land and at sea is that the fungi, especially yeasts and filamentous types are generally considered to play a much more important role in soil than in water. However, it should be noted that the previous authors [13] could not demonstrate a significant increase in the number of hydrocarbon utilizing fungi, although this may be a limitation of the method used. The question of fungal degradation of oil in the marine environment has recently been reviewed [15].

In addition to the catastrophic spillages mentioned, there is also considerable endemic pollution caused by various procedures such as the loading and unloading of oil tankers. Dudley [16] has estimated that 0.0001% of the oil passing through Milford Haven is lost.

Other forms of hydrocarbon pollution occurring at sea will include oil from tankers flushing storage tanks, industrial wastes from the manufacture of various oils and soaps and the long chain hydrocarbon portions of certain detergents and industrial cleaning and sterilizing agents.

Most serious oil pollution occurs near the coast and requires immediate intervention to avoid gross pollution of the intertidal zone, which is one of intense biological activity [17,18,19,20,21] and probably where the most destructive effects occur.

The corrective measures taken may be of several types. There is the purely physical treatment such as the containing of the oil by floating booms, followed by the addition of an absorbent such as hay or straw, which has been used on enclosed waters for small amounts of oil.

Chemicals added to disperse oil slicks may be of two types: those causing sinking and those causing dispersal.

Oil which has been weathered tends to sink and this can be accelerated by the use of chalk or sand coated with silicones or long chain amines. Ideally the additive adheres to the oil, sinks it, and does not release it so that the oil is completely immobilized [22]. Sinking is now only used as a last resort, as recent work has shown that the oil degrading activities of micro-organisms in the water is greater than those of the sediment [23].

Oil can also be dispersed by the use of surfactants whose role is helped by wind and wave action, but the two primary requirements are that the surfactant should be non-toxic and biodegradable [24].

Crude oil spilt into the sea undergoes a variety of physical and chemical changes. This means that no two oil spillages will be identical in terms of microbial substrate. Evaporation is obviously a highly variable factor and the volatile components lost will depend not only on the type of oil but also on the weather conditions. The effect of ambient temperature on volatile compounds is obvious, but factors such as the rate at which the oil spreads will also affect the rate of loss. Dean [25] considers that approximately two thirds of Nigerian crude oil evaporates after a few days at sea, whereas only about forty per cent of Venezuelen crude will evaporate under the same conditions. Fractions boiling at less than $350^{\circ}C$ will evaporate within approximately one week.

The temperature of the water will also affect the rate of microbial metabolism. In the temperature range $0-40^{\circ}C$ most enzymes (biological catalysts), approximately double their reaction rate for every rise of $10^{\circ}C$.

Degradation of oil may take place aided by physical factors such as u.v. light which causes [25] the formation of peroxide which can be degraded or dimerized to complex molecules [26]. Autoxidation also takes place, the rate of this depending upon the presence of various chemicals; cations generally accelerate the process whereas phenolics and sulphur containing compounds act as inhibitors [27,28,29].

The rate of biodegradation will depend very considerably on the physical state of the oil as well as its chemical state. The influence of temperature on enzyme catalysis has been mentioned but Nyns [30] has suggested that the rate limiting factor in the oxidation of hydrocarbons may be their solubility.

The solubility of hydrocarbons in aqueous solution is low and decreases as the chain length and molecular weight increase. The solubility of hexane (mol.wt. 80) has been calculated at 138 p.p.m. and that of tetradecane (mol.wt. 198) at 2.0×10^{-6} p.p.m. [31]. Hydrocarbons are also less soluble in sea water than distilled water so that the soluble hydrocarbon substrate available for microbial attack is very limited. This probably accounts for the fact that the growth rate of bacteria growing on hydrocarbons is related to the solubility of the hydrocarbon.

The degree of dispersal of the oil will also affect its rate of oxidation, a thin film offering a very large surface area for bacterial attack and easy access to oxygen and dissolved nutrients. Conversely oil micelles cannot exchange nutrients as easily and the possibility of the accumulation of toxic wastes arises [32]. Oxygen is unlikely to be the rate limiting factor as it readily dissolves in oil [30] and the hydrocarbon concentration is low. It is probable that the rate limiting factors of growth are supplies of nitrate and phosphate, and a recent paper [33] described the use of a slow release fertilizer containing $MgNH_4PO_4$ carried on a paraffin support. The authors show that the use of such fertilizers enhances oil degradation by a factor of over 50% in a given time period.

It should be noted however that several authors have suggested that some micro-organisms capable of oxidizing hydrocarbons may also be able to fix atmospheric nitrogen [34,35].

The biodegradation of oil and hydrocarbons has been known for a considerable period of time and lists of bacteria and fungi capable of degrading hydrocarbons are given by a number of authors [15,36,37,38]. The biodegradation of oil in the marine environment has recently been considered by Floodgate [39].

Much of the research work on biodegradation has been carried out using pure hydrocarbon substrates and pure cultures of micro-organisms, and as such it is not applicable to the situation found in the natural environment. Differences have been reported in the biodegradability of crude and fuel oils [40] and such differences may well be accentuated by the use of pure substrates.

Perry and Scheld [41] attempted a more ecological approach when they isolated hydrocarbon utilizers from soil. The conclusion they reached was

been discredited to some extent as the very low K_m for many oxygenases [51]
make it possible for chemically undetectable levels of oxygen to allow
oxidation of the substrate alkane.

A number of pathways proposed for the aerobic degradation of hydrocarbons have been reviewed [52,53]. There is a considerable amount of substrate specificity towards the chain length of the substrate. There is a
dividing line at C_{10} with the majority of micro-organisms favouring the
shorter alkanes. Certain species of bacteria show a total specificity
towards methane and the special problems relating to this group have been
reviewed by Quayle [54].

In the predominant aerobic pathway of alkane degradation, oxidation
occurs at the terminal carbon atom of the alkane. Enzyme systems isolated
from <u>Pseudomonas</u> are capable of oxidizing the alkane to alcohol [55], the
alcohol to the aldehyde [56] and the aldehyde to the corresponding fatty
acid [57] as shown in equations 1-3.

$$R-CH_2-CH_2-CH_3 \xrightarrow{[o]} R-CH_2-CH_2-CH_2OH \tag{1}$$

$$R-CH_2-CH_2-CH_2OH \xrightarrow{[o]} R-CH_2-CH_2-CHO + H_2O \tag{2}$$

$$R-CH_2-CH_2-CHO \xrightarrow{[o]} R-CH_2-CH_2-COOH \tag{3}$$

Once the fatty acid has been synthesized it is broken down by the
β-oxidative degradation pathway shown in equations 4-8. These reactions
involve the synthesis of an acyl CoA ester which is shortened by two carbon
atoms at a time on each occasion it enters the β-oxidative cycle. The long
chain fatty acid is eventually completely broken down to acetyl CoA [58].

$$R-CH_2-CH_2-COOH + ATP + CoA \longrightarrow R-CH_2-CH_2-CO-CoA + AMP + PP_i \tag{4}$$

$$R-CH_2-CH_2-CO-CoA + FAD \longrightarrow R-CH=CH-CO-CoA + FADH_2 \tag{5}$$

$$R-CH=CH-CO-CoA + H_2O \longrightarrow R-CH(OH)-CH_2-CO-CoA \tag{6}$$

$$R-CH(OH)-CH_2-CO-CoA + NAD^+ \longrightarrow R-CO-CH_2-CO-CoA + NADH + H^+ \qquad (7)$$

$$R-CO-CH_2-CO-CoA + CoA \longrightarrow R-CO-CoA + CH_3-CO-CoA \qquad (8)$$

Other bacteria beside those already mentioned have been shown to oxidize hydrocarbons [59,60,61,62].

A number of fungi oxidize n-alkanes, the most important group being the yeast Candida. Several authors have obtained cell-free extracts of yeast capable of converting alkanes to the corresponding alcohols and aldehydes [63,64], whilst Ratledge [65] has shown that the fatty acid pattern of yeasts grown on alkanes reflects the alkane used as substrate.

Alkanes are not the only components of oil, but the degradation of aromatics is dealt with elsewhere.

One factor not as yet mentioned, but assuming increasing importance, is that of cometabolism and commensalism. Cometabolism has been described by Leadbetter and Foster [66] who showed that methane was the only hydrocarbon capable of supporting the growth of Pseudomonas methanica. If however ethane, propane or butane were added, then these compounds were partially oxidized to acetic, propionic and butyric acids respectively. Jensen [67] suggested the use of the term cometabolism to describe the process whereby an organism oxidized a non-growth supporting compound in the presence of a growth supporting compound. A number of examples are now known including a variety of dehalogenation reactions carried out by both bacteria and fungi. It has been shown that it is possible to select microorganisms capable of the cometabolic degradation of organic pollutants by applying a biodegradable analogue of the pollutant to the ecosystem [68].

A recent article by Beam and Perry [69] discusses the microbial degradation of cyclohexane in marine mud and concludes that cometabolism is partially responsible, but that a commensal attack by two or more microorganisms working in concert may also be involved.

A review by Horvath [68] summarizes much of the present knowledge and concludes that the problem of molecular recalcitrance or non-biodegradability may well have to be reconsidered.

The growth of micro-organisms on oil has so far been considered as a desirable asset in reducing oil pollution, but there are a large number of situations in which the growth of micro-organisms on oil is definitely nondesirable [70].

Petroleum products are invariably involved in the production of sheet steel and aluminium, not only as lubricating oils but also for cooling purposes [71]. The amount of oil is frequently large (up to 100,000 gallons) and conditions often favour the rapid growth of micro-organisms. These organisms grow actively only in the aqueous phase and if carried into the oil phase they only survive for a few hours [72]. In the aqueous phase however, populations of 10^9 bacteria per ml are easily reached and the bacteria may contribute up to 5% of the weight of the aqueous layer. The level reached is frequently sufficient to clog filters, as was seen by the authors in a case of Cladosporium resinae developing in a metal finishing process.

The chemical changes occurring in the oil discussed earlier can be followed in terms of progressive failure in the engineering process. When oil is used in a factory it is frequently contaminated to a high level with micro-organisms from the slime left by the previous process. In continuously used well aerated systems kept at temperatures between 20-40°C bacterial growth is rapid and species of Pseudomonas, Alcaligenes and Achromobacter dominate. Pseudomonas aeruginosa has been shown as a contaminant in hard water regions [73]. If the system is poorly aerated and used intermittently, then anaerobic organisms particularly Clostridium nigrificans and facultative anaerobes will be present. Rolling oil emulsions kept at elevated temperatures will suppress the growth of mesophilic organisms and allow thermophiles such as species of Nocardia to dominate.

In addition to the deleterious effects of these organisms on the oil, there is also the problem of health hazards, especially in equipment that could cause aerosols. Pseudomonas species have been implicated in septic conditions following a wound and also in eye infections, whilst Clostridia species are notorious for their ability to contaminate piercing wounds.

The microbiology of any system depends upon a number of factors which include the rate of aeration, temperature, pH, availability of nutrients and the presence of inhibitors. In normal emulsions there is an excess of carbon for microbial nutrition but nitrogen, phosphate, sulphate and trace elements may be deficient. In most works these can be provided by debris such as rat urine and faeces.

The microbial activity can be seen in many forms, the most obvious being the appearance of slimes and sediments if they are allowed to accumulate. These can be removed in centrifuge or paper swarf separators. There may be objectionable smells produced due to degradation products and these are frequently associated with a darkening in colour.

The chemical changes occurring can affect the lubricating properties and viscosity of the oil and may also alter any additives incorporated into the oil to increase film strength, induce 'bite', inhibit corrosion or stabilize emulsions. Bacterial activity may alter droplet size distribution of the oil in water and eventually lead to separation of the oil. This can be related to a number of known defects in aluminium products [74] and is a cause of premature emulsion failure.

High microbial activity in oil emulsions may cause poor surface finish during machining and rolling and roll-slip in rolling steel and aluminium. If the micro-organisms grow on the sheet or are deposited on it, on annealing they are converted to a hygroscopic ash which may be a focus for corrosion [75]. In some cases if hydrogen sulphide producing anaerobes survive on the sheet, then a grey sulphide stain may develop prior to annealing. There may also be fouling of grinding wheels, wheel burn and high wheel dressing costs with rapid corrosion of the metal after machining. High solid loads may occur on filters and clarifiers with reduced filter efficiency and reduced oil flow and pressure.

It may thus be seen that high microbial activity in oil emulsions can have many economic effects. These include short oil life with costly oil changes, loss of production time during oil changes, excessive topping up to maintain emulsion strength and faulty finished products possibly leading to premature failure.

These problems may be minimized by considering the oil formulation to ensure that the components are not an ideal substrate for microbial growth. Physical and/or chemical methods may be used to inhibit growth, for example, the addition of antibacterial agents to reduce or eliminate an infection. Any chemical additives must however be screened to ensure that they do not adversely affect the finished product [76]. It is also beneficial to carry out simple routine tests to determine the microbial level and monitor the efficiency of any antimicrobial agent which has been added [77].

REFERENCES

1 P.E. Kolattukudy, Science, 159(1968)498.
2 P.E. Kolattukudy, Ann. Revs. Plant Physiol., 21(1970)163.
3 G. Egglinton and R.J. Hamilton, Science, 156(1967)1322.

4   T. Kaneda, Phytochemistry, 8(1969)2039.

5   R.C. Clark and M. Blumer, Limnol. Oceanog., 12(1967)79.

6   J. Oro, T.G. Tornabene, D.W. Nooner and E. Gelpi, J. Bact., 93(1967) 1811.

7   J. Han and M. Calvin, Proc. Nat. Acad. Sci. U.S., 64(1969)436.

8   T. Kaneda, Biochemistry, 7(1968)1194.

9   J. Avigan and M. Blumer, J. Lipid Res., 9(1968)350.

10 D.E. Hughes and P.M. McKenzie, Proc. Roy. Soc., London B, 189(1975)375.

11 G.D. Floodgate, in J.M. Aderson and A. Macfadyen (Editors), The Role of Terrestrial and Aquatic Organisms in Decomposition Processes, Blackwell, Oxford, 1976, Ch. 9, p. 223.

12 W.B. Travers and P.R. Luney, Science, 194(1976)791.

13 R.L. Raymond, J.O. Hudson and V.W. Jamison, Appl. and Environ. Microbiol., 31(1976)522.

14 R.M. Atlas and M. Burdosh, in J.M. Sharpley and A.M. Kaplan (Editors), Proc. 3rd Int. Biodegradation Symp., Applied Science Publishers, 1976, p. 79.

15 D.G. Ahearn and S.P. Meyers, in E.B. Gareth Jones (Editor), Recent Advances in Aquatic Mycology, Elek Science, 1976, Ch. 4, p. 125.

16 G. Dudley, in J.D. Carthy and D.R. Arthur (Editors), The Biological Effects of Oil Pollution on Littoral Communities, Field Studies Council, 1968, p. 21.

17 D.H. Dalby, ibid. p. 31.

18 D.S. Ranwell, ibid. p. 39.

19 D.J. Bellamy and A. Whittick, ibid. p. 49.

20 A.D. Boney, ibid. p. 55.

21 A. Nelson-Smith, ibid. p. 73.

22 N. Pilpel, Research (Lond.), 7(1954)301.

23 J.D. Walker, R.R. Colwell and L. Petrakis, Applied Microbiol., 30 (1975)1036.

24 W. Gunkel, in J.D. Carthy and D.R. Arthur (Editors), The Biological Effects of Oil Pollution on Littoral Communities, Field Studies Council, 1968, p. 151.

25 R.A. Dean, ibid. p. 1.

26 M. Freegarde, C.G. Hatchard and C.A. Parker, Lab. Practice, 20(1971)35.

27 C.E. Frank, Chem. Revs., 46(1950)155.

28 R.M. Atlas and R. Baritha, Canad. J. Microbiol., 18(1972)1851.

29 A.C. Nixon, in W.O. Lundberg (Editor), Autoxidation and Antioxidants, Interscience, Vol. 2, 1962.

30 E.J. Nyns, Rev. Questions Sci., 28(1967)189.

31 M.J. Johnson, Chem. Ind. (London), 36(1964)1532.

32 C.E. Zobell, Int. J. Air Water Pollut., 7(1963)173.

33 R. Olivieri, P. Bacchin, A. Robertiello, N. Oddo, L. Degen and A. Tonolo, App. and Environ. Microbiol., 31(1976)629.

34 H.J. Harper, Soil Sci., 48(1939)461.

35 V.F. Coty, Biotechnol. Bioeng., 9(1967)25.

36 G.W. Fuhs, Arch. Mikrobiol., 39(1961)374.

37 A.P. Ponsford, Mon. Bull. Brit. Coal Utd. Res. Assoc., 30(1966)41.

38 M.J. Klug and A.J. Markovetz, in A.H. Rose and J.F. Wilkinson (Editors), Advances in Microbial Physiology, Vol. 5, 1971, Ch. 1, p. 1.

39 G.D. Floodgate, in J.M. Sharpley and A.M. Kaplan (Editors), Proc. 3rd Int. Biodegradation Symp., Applied Science Publishers, 1976, p. 87.

40 J.D. Walker, L. Petrakis and R.R. Colwell, Canad. J. Microbiol., 22 (1976)598.

41 J.J. Perry and H.W. Scheld, Canad. J. Microbiol., 14(1968)403.

42 D.C. Gordon, P.D. Keizer, W.R. Hardstaff and D.G. Aldous, Environ. Sci. Tech., 10(1976)580.

43 J.B. Davis, Petroleum Microbiology, Elsevier, Amsterdam, 1967.

44 A. Champagnot, in B. Spencer (Editor), Industrial Aspects of Biochemistry. Proc. Fed. European Biochem. Soc., North Holland/ American Elsevier, Vol. 30, 1974, p. 347.

45 Protein-Calorie Advisory Group of the United Nations System, Vol. 6, Part 2, 1976.

46 A. Spicer, Br. Med. Bull., 31(1975)220.

47 R.W. Hansen and R.E. Kallio, Science, 125(1957)1198.

48 D.G. Ahearn and S.P. Meyers, in A.H. Walters and H. van der Plas (Editors), Biodeterioration of Materials, Vol. 2, Applied Science, London, 1972, p. 12.

49 J.B. Davis and H.F. Yarbrough, Chem. Geol., 1(1966)137.

50 R.W. Traxler and J.M. Bernard, Int. Biodeterior. Bull., 5(1969)21.

51 J.A. Davies and D.E. Hughes, in J.D. Carthy and D.R. Arthur (Editors), The Biological Effects of Oil Pollution on Littoral Communities, Field Studies Council, 1968, p. 139.

52 D.G. Floodgate, in R. Mitchell (Editor), Water Pollution Microbiology, Wiley-Interscience, New York, 1972, p. 153.

53 C.W. Bird and P.M. Molton, in F.D. Gunstone (Editor), Topics in Lipid Biochemistry, Vol. 3, 1974, p. 125.

54 J.R. Quayle, in A.H. Rose and J.F. Wilkinson (Editors), Advances in Microbial Physiology, Vol. 7, 1972, p. 119.

55 R.K. Gholson, J.N. Baptist and M.J. Coon, Biochem., 2(1963)1155.

56 A.C. van der Linden and R. Huybregtse, Ant. van Leeuwenhoek, 35(1969) 344.

57 M.T. Heydeman and E. Azoulay, Biochim. Biophys. Acta, 77(1963)545.

58 A.L. Lehninger, Biochemistry, Worth Publishers, New York, 2nd ed., 1975, Ch. 20, p. 543.

59 J.J. Perry, Ant. van Leeuwenhoek, 31(1968)45.

60 K.R. Dunlap and J.J. Perry, J. Bact., 96(1968)318.

61 W.R. Finnerty, E. Hawtrey and R.E. Kallio, Z. Allgem. Mikrobiol., 2 (1962)169.

62 K.M. Fredricks, Ant. van Leeuwenhoek, 33(1967)41.

63 H. Iizuka, M. Iida, Y. Unami and Y. Hoshino, Z. Allgem. Mikrobiol., 8 (1968)145.

64 H. Iizuka, M. Iida and S. Toyoda, Z. Allgem. Mikrobiol., 6(1966)335.

65 C. Ratledge, Biotechnol. Bioeng., 10(1968)511.

66 E.R. Leadbetter and J.W. Foster, Arch. Biochem. Biophys., 82(1959)491.

67 H.L. Jensen, Acta Agri. Scand., 13(1963)404.

68 R.S. Horvath, Bact. Revs., 36(1972)146.

69 H.W. Beam and J.J. Perry, J. Gen. Microbiol., 82(1974)163.

70 E.C. Hill, in J.M. Sharpley and A.M. Kaplan (Editors), Proc. 3rd Int. Biodegradation Symp., Applied Science Publishers, 1976, p. 243.

71 D. Byrom and E.C. Hill, in P. Hepple (Editor), Microbiol. Proc. Conf., 1971, p. 42.

72 E.C. Hill, D.A. Evans and I. Davies, J. Inst. Petroleum, 53(1967)280.

73 E.V. Feisal and E.O. Bennet, J. Appl. Bact., 24(1961)125.

74 E.C. Hill, Engineering, 203(1967)983.

75 E.C. Hill and I.O. Pemberthy, Metals Mat., 2(1968)359.

76 H.U. Kleist and E.C. Hill, Aluminium (Dusseldorf), 48(1972)606.

77 E.C. Hill, I. Davies, J.A.V. Pritchard and D. Byrom, J. Inst. Petroleum, 53(1967)275.

BIODEGRADATION

Large amounts of complex organic molecules containing aromatic, heterocyclic or alicyclic ring systems are synthesized naturally every year [1,2]. Complex molecules produced biosynthetically have always been capable of degradation by bacterial and microbial enzymes, even though the time required for total degradation may have been considerable.

These naturally occurring compounds have been supplemented in recent decades by a large number of synthetic chemicals used for a wide variety of purposes. Synthetic chemicals frequently present a new problem in bio-degradation, as many of them contain halogenated groups and are thus novel substrates to the microbial degradative enzymes. However, recent work suggests that in some cases, it is the amount of synthetic compound involved rather than its structural stability which is the cause of pollution.

The problem of molecular recalcitrance is one which is undergoing considerable re-examination at the present moment, as the significance of biochemical studies carried out using pure cultures and pure substrates is highly questionable. Water, soil and sewage are extremely complex environments, biologically, chemically and physically. Large numbers of different types of organisms may be present, displaying a wide range of degradative abilities. Partial degradation of compounds may occur under the influence of physical factors such as u.v. light. The presence of various organic or inorganic chemicals is known to catalyse non-enzymic reactions and accelerate or retard enzymic reactions. The type and size of particles, especially in soil and mud, may cause widely fluctuating localized changes in substrate concentrations, pH, mineral and water content. The presence or absence of predators and parasites may also cause microbial populations to fluctuate over very wide ranges.

This wide range of parameters may mean that degradative pathways found in one ecosystem may not be the same as those found in another complex eco-system.

Aromatic Degradation

One of the most important sources of organic carbon is lignin which is found to a large extent in wood and is a polymer of the three aromatic compounds shown.

194

Fig.1

a)

COOH — CHOH → COOH — CO → CHO → COOH

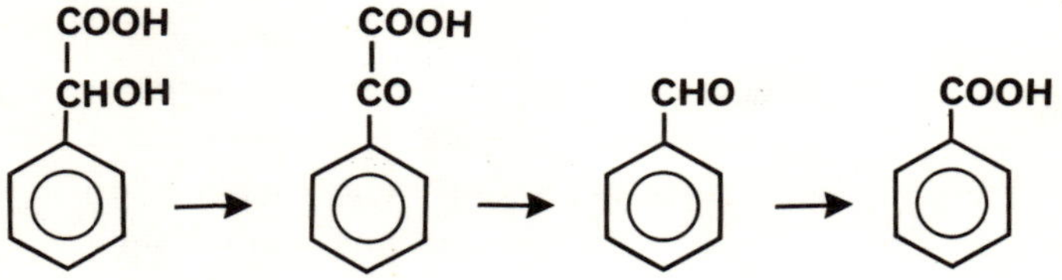

Mandelic acid   Benzoic acid

b)

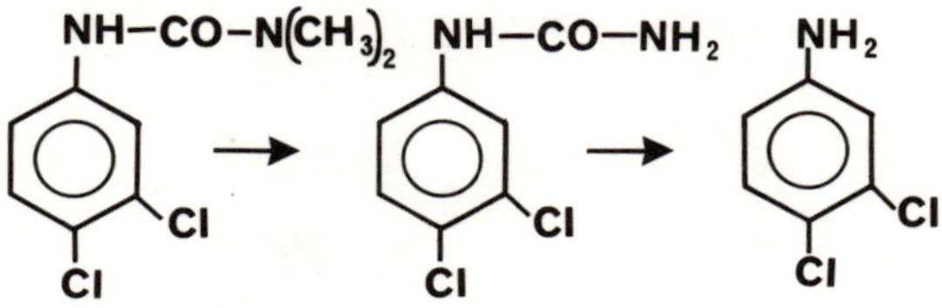

NH—CO—N(CH₃)₂ → NH—CO—NH₂ → NH₂

Cl, Cl   Cl, Cl   Cl, Cl

Diuron   3,4-Dichloroaniline

c)

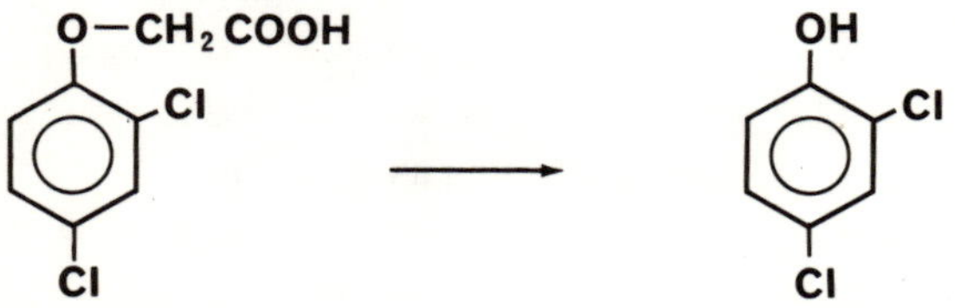

O—CH₂COOH → OH

Cl, Cl   Cl, Cl

2,4-Dichlorophenoxy-   2,4-Dichlorophenol
acetic acid (2,4-D)

d)

CH₃ → CH₂OH → CHO → COOH

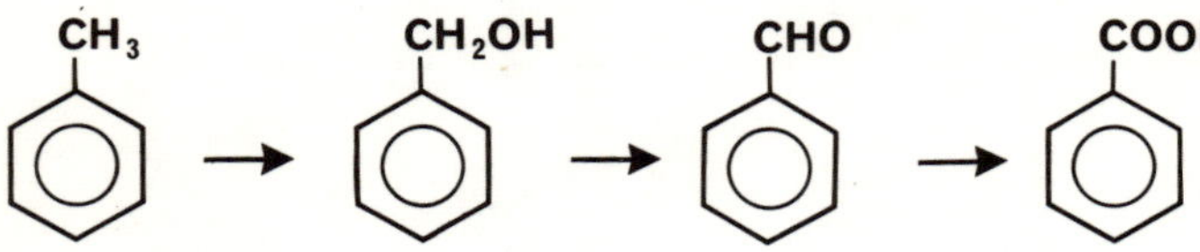

Toluene   Benzoic acid

p-Coumaryl alcohol   Coniferyl alcohol   Sinapyl alcohol

    Smaller amounts of other aromatic chemicals are also found, including a number of naturally occurring complex polycyclic compounds [2].

    Much of the work on the degradation of aromatic compounds has been carried out using the bacterial genus Pseudomonas and this has been recently reviewed [3,4]. Pseudomonas species are probably the most bio-chemically versatile group of bacteria, much of this property being due to the possession of various types of plasmid (see later).

    The first sequence of reactions carried out is usually the removal of any side chain that may be present (Fig. 1), although there are exceptions to this.

    Once the side chain has been removed, the next requirement is for hydroxylation of the aromatic ring. This is carried out by a group of enzymes known as oxygenases, which occur at a number of points in the degradative pathways and can be split into two types.

    The first group are known by several names, mono-oxygenases, hydroxylases and mixed function oxidases. These enzymes add one atom of oxygen to a substrate molecule and reduce the second atom of oxygen to water by means of a reduced enzymic cofactor as shown in reaction 1.

$$\text{(1)}$$

Fig. 2

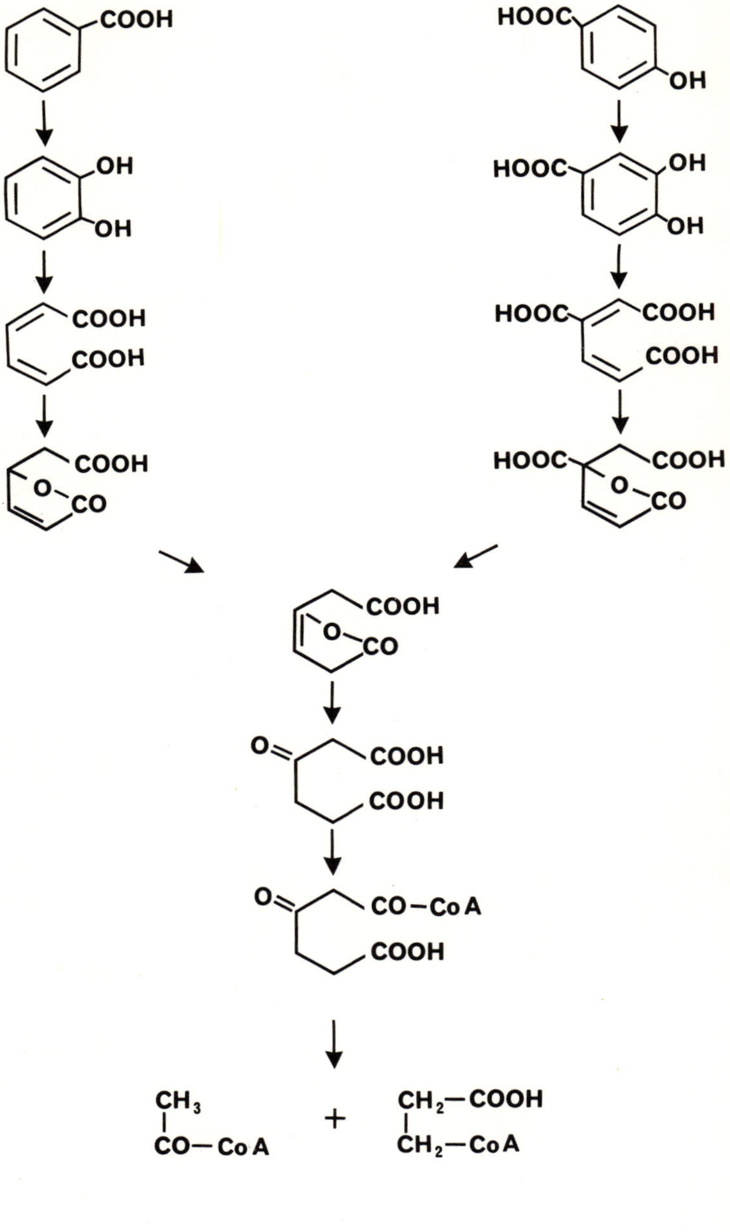
COOH
OH
OH
COOH
COOH
COOH
O
CO
HOOC
OH
HOOC
OH
OH
HOOC
COOH
COOH
HOOC
COOH
O
CO
COOH
O
CO
O
COOH
COOH
O
CO—CoA
COOH
CH3
CO—CoA
+
CH2—COOH
CH2—CoA

The second group of oxygenases are the dioxygenases which add two atoms of oxygen to a substrate molecule and do not utilize a reduced co-enzyme with the concomitant formation of water. An example of this is the dihydroxylation of benzoic acid to catechol (reaction 2) and the ortho cleavage of catechol shown in reaction 3 to give cis-cis muconic acid.

(2)

(3)

The substrate specificity of many dioxygenases is limited and Dagley [1] lists a number of such enzymes capable of oxidizing a wide variety of aromatic substrates.

The initial reaction in the fission of the aromatic ring is the introduction of two hydroxyl groups into the aromatic nucleus ortho to each other. In the case of an aromatic ring already containing a hydroxyl substitution a second hydroxyl group is introduced ortho to the first. The types of reactions found are shown in reactions 1 and 2.

After this, further degradation of catechol or the substituted derivative of catechol may take place by one of two pathways known as ortho or meta cleavage.

The first type of cleavage known as ortho cleavage involves reaction 3 and gives rise to β-keto adipic acid which is ultimately split to acetyl coenzyme A and succinyl coenzyme A as shown in Fig. 2.

The second type of aromatic cleavage known as meta cleavage or the α-keto acid pathway also takes place by means of a dioxygenase (reaction 4) to give 2-hydroxy-muconic semialdehyde. In this case the ultimate products from catechol will be acetaldehyde and pyruvic acid as shown in Fig. 3.

Fig. 3

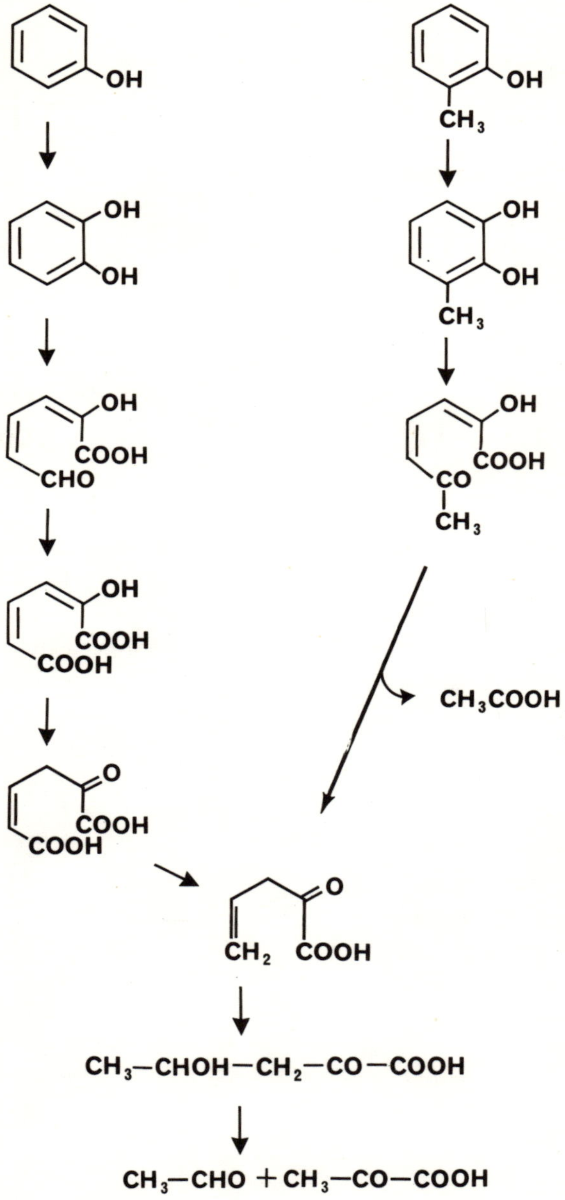

$$\text{(4)}$$

It can be seen from reactions 3 and 4 that the same compound may be subjected to different pathways in different organisms. Evidence exists to show that in certain cases catechol may be cleaved by either pathway in the same organism as shown in reaction 5 [5].

$$\text{(5)}$$

Davies and Evans [6] showed that one strain of Pseudomonas putida would split catechol derived from benzoate by the ortho pathway, whereas catechol derived from naphthalene would be split by the meta route. This is obviously a problem of enzyme control which is discussed later.

The enzymes catalysing these degradative pathways are highly specific. The breakdown of benzoic acid and p-hydroxybenzoic acid shown in Fig. 2 is catalysed by different enzymes until they reach the common intermediate β-keto adipic acid [7,8].

The study of aromatic degradation shows that the breakdown of many of these compounds is subject to tightly controlled regulations [3,4,7,8,9,10]. This control can take place by means of enzyme induction and enzyme repression.

Many enzymes are present in the cell in trace amounts under normal metabolic conditions, but in the presence of specific enzyme inducers, the

synthesis of permease (transport enzymes) and thus stopping the entry of growth substrates into the cell [3].

A study of the repression system (Fig. 4) shows that broadly speaking an organism will use the most easily available source of carbon and energy first, that is succinate and acetate. Mandelic acid will not be utilized until the levels of benzoate, catechol and succinate have fallen sufficiently to allow the concentration of mandelate to overcome the inhibition. This point is of obvious importance in a study of pollutant degradation in a complex environment, which may have access to a large number of substrates giving rise to compounds such as catechol or succinate. It should be stressed however that very little work has been carried out on the control of enzymes responsible for the degradation of specific pesticides in the natural environment.

There is the possibility that bacteria will mutate to give cells which are constitutive for the degradation enzymes. When this happens the degradative enzymes are always present at high levels, that is they are no longer induced. It is relatively easy to produce this type of change under laboratory conditions, and Parker and Ornston [14] have reported the constitutive synthesis of enzymes degrading protocatecheuic acid in Pseudomonas putida. The importance of such mutations in the natural environment is not well documented, but Dagley [1] considers the possible evolution of degradative pathways and the matter is considered in greater detail later.

The induction system has been shown to be different in other species of Pseudomonas.

Murray and Williams [5] discuss the role of catechol as an enzyme inducer in Pseudomonas putida which can cleave catechol by either the ortho or meta pathway (reaction 5). The organism metabolizes benzoate by the ortho route and phenol by the meta route.

The explanation is that phenol acts as inducer for the entire suite of meta degradation enzymes (Fig. 3), whereas benzoate only induces the enzyme benzoate dihydroxylase to synthesize catechol. Small amounts of catechol are then metabolized to form cis-cis muconate which induces the synthesis of the next group of ortho degradative enzymes.

The ortho pathway can thus only be induced in the absence of an inducer of the meta pathway.

The pathways already considered are applicable to a large number of naturally occurring and synthetic aromatic compounds. The breakdown of

(Continued on p. 205)

Fig 5

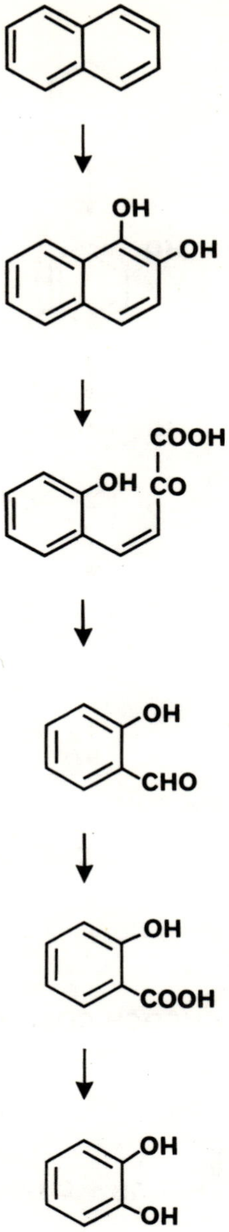

Fig. 6

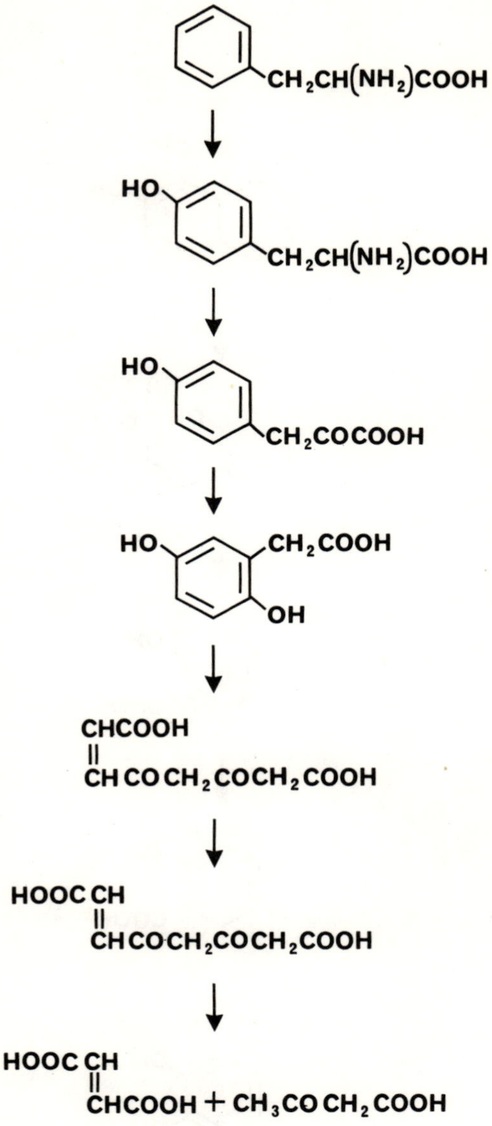
CH2CH(NH2)COOH
HO
CH2CH(NH2)COOH
HO
CH2COCOOH
HO
CH2COOH
OH
CHCOOH
CH CO CH2 COCH2 COOH
HOOC CH
CHCOCH2COCH2COOH
HOOC CH
CHCOOH + CH3CO CH2 COOH

polyhydroxy aromatic compounds is considered by Chapman and Ribbons [15].
Polycyclic aromatics will be broken down by a variety of oxygenases
extending this system. Naphthalene for example will be oxidized as shown
in Fig. 5, whilst phenanthrene will be oxidized in a similar manner to give
phenanthrene 1:2 diol which will then be degraded to naphthalene 1:2 diol.
The degradation of polycyclic hydrocarbons is considered by Gibson [16].

The degradation of the aromatic amino acids, tyrosine and phenylalanine
follows a somewhat different pathway, although these two compounds can
hardly be considered as pollutants. Their degradation is included for the
sake of completeness. The degradation pathway is shown in Fig. 6. A
number of substituted gentisic acids can be broken down by the same route.
The initial reaction is a p-hydroxylation of the type shown in reaction 1.
The unusual step is carried out by the enzyme converting p-hydroxyphenyl-
pyruvic acid to homogentisic acid. This copper containing enzyme carries
out four functions, decarboxylation, oxidation of the keto group, hydroxy-
lation of the aromatic ring and migration of the side chain.

Organisms shown to be capable of degrading the aromatic ring include a
number of bacteria such as Pseudomonas, Achromobacter, Bacillus and
Arthrobacter, whilst amongst the fungi Penicillium, Aspergillus and
Fusarium are frequently mentioned in the literature. Very little work
appears to have been carried out on the possibility of algae being
involved in biodegradation.

A study of the genetics of aromatic catabolism in the bacterial genus
Pseudomonas has been carried out over the last few years. Results
obtained with Pseudomonas shows that in certain species of this genus some
of the degradative pathways are coded for by the presence of plasmids.
These plasmids are extra chromasomal deoxyribonucleic acid (DNA), that is
DNA not associated with the nucleus of the bacterial cell.

The degradative pathways coded for have been shown to be salicylate
[17] m- and p-toluate (methyl benzoate) [18], toluene and xylene [19,20],
naphthalene [21] and octane [22].

Worsey and Williams [20] have shown that in Ps. putida the TOL plasmid
is responsible for the synthesis of enzymes degrading toluene as shown in
Figs. 1(d) and 3. If these cells were treated with benzoate they metabolized it via the meta cleavage pathway (Fig. 3). Cured cells (those
which have lost the plasmid) occur spontaneously at a fairly high frequency,
and Williams and Worsey [19] have shown that cured cells can no longer
metabolize toluene or xylene and metabolize benzoate by the chromosomally

coded _ortho_ pathway (Fig. 2).

The genetics of degradation pathways in pseudomonas species has been recently reviewed by Whellis [23].

The suggestion has been made that plasmids of this type could be deliberately introduced into Pseudomonas which could then be sprayed onto oil slicks, or into polluted ecosystems causing accelerated biodegradation. However, early results suggest that there may be an upper limit to the number of plasmids any one bacterium can carry, and the presence of certain plasmids appears to be incompatible with the presence of others. In laboratory experiments the rate of spontaneous curing in the absence of the substrate is also comparatively high, which gives rise to questions about the stability of such organisms in the natural environment.

Probably one of the more important ecological implications of this work is that it is no longer possible to consider one bacterial species in genetic isolation to another, and in a mixed bacterial population in a complex environment, the gene pool of any one species of Pseudomonas can no longer be regarded as sacrosanct.

Heterocyclic and Alicyclic Degradation

The degradation of some naturally occurring heterocyclic ring systems such as the purines and pyrimidines is well documented and may be found in many standard biochemistry texts [24].

The five membered heterocyclic rings of pyrrole, furan and thiophen are frequently considered together because of their superficial similarity.

The degradation of furan-2-carboxylic acid is shown in Fig. 7, and the biodegradation of thiophen which is a common component of tar and oil is thought to be similar [25,26].

Fig. 7

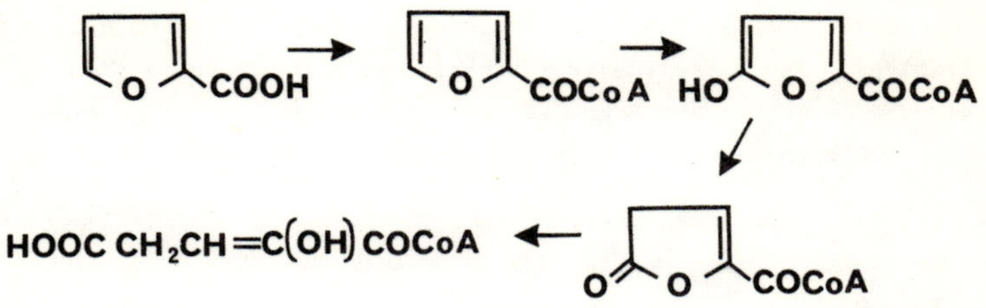

The pyridine ring is also found in naturally occurring compounds such as the nicotinamide containing coenzymes and the alkaloid nicotine. Pyridine derivatives are also found in many industrial effluents and pyridine derivatives are also used as herbicides, for example, the compounds paraquat and diquat. The degradation of the pyridine ring is shown in Fig. 8.

Fig. 8

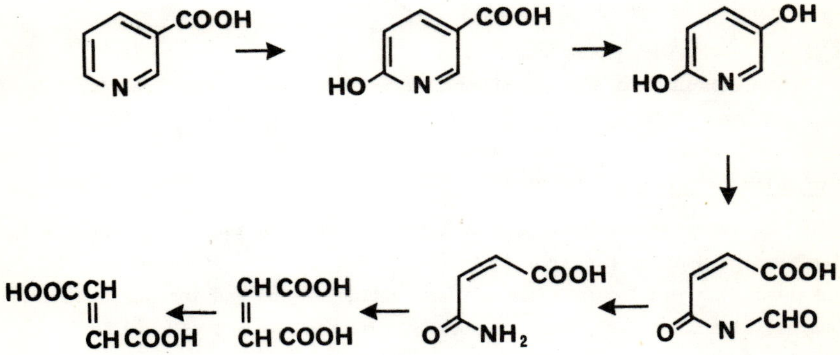

The cleavage of alicyclic rings has not been studied as intensely as the cleavage of the aromatic ring. Cyclohexane is comparatively resistant to microbial attack, and persists in marine situations for long periods.

Several pathways have been proposed for degradation of cyclohexane derivatives. These include aromatisation of the alicyclic ring to the dihydroxy-derivative and degradation by either meta or ortho cleavage, or alternatively the conversion of the alicyclic ring to a lactone as shown in Fig. 9.

Fig. 9

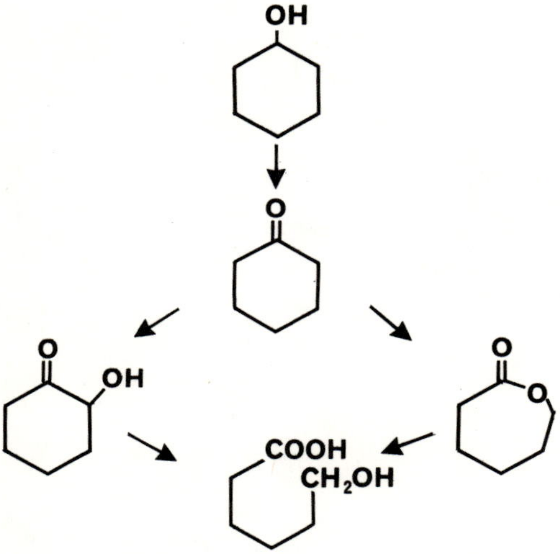

The degradation of alicyclics is considered briefly by Trudgill [26] and reviewed by Donaghue et al. [27].

Synthetic Detergents

The increased domestic use of synthetic detergents started after 1949. The typical washing powder contains about 20% w/w of surface-active material, the remainder is present as "builders" and can include polyphosphates, silicates, sulphates, carboxymethylcellulose and alkonamides. These materials are added to increase the efficiency of the washing powder, stabilise the foam and prevent re-deposition of dirt. Optical whiteners, colour and perfume are normally added. Synthetic detergents are also used in washing-up liquids and scouring powders. The surface-active material is usually a mixture of alkyl sulphonates $CH_3(CH_2)_n-C_6H_4SO_3Na$ where (n = 9-13).

The presence of synthetic detergents in sewage produces banks of foam at aeration points, and where the effluent is discharged into the river. This foam can be controlled by use of an antifoam agent.

Many of the problems associated with the discharge of domestic synthetic detergents result from an incomplete degradation of the surface-active material during normal sewage treatment processes. The ease of breakdown depends on the exact configuration of the alkyl chain. If the alkyl benzene sulphonate has a straight chain of 10 carbon atoms with the benzene ring attached to a terminal carbon atom, it is readily degraded. When a branched carbon chain is present this is resistant to microbial attack.

Biodegradability can be tested in a variety of ways [28,29,30] usually in aerobic environments. In the case of anionic detergents, a simple screening test against pure cultures of selected bacteria and fungi can result in an underestimate of biodegradability. Tests like the O.E.C.D. procedure [31] are widely applied to estimate the biodegradability of detergents under simulated activated sludge conditions.

The presence of linear alkyl sulphonates at 720 mg/l can alter the performance of activated sludge causing inhibition of biodegradation [32]. A substantial portion of the detergent, however, is absorbed onto sludge flocs and eliminated in this way. The branched chain nonyl phenol ethoxylates at 5.5 mg/l are biodegraded with an excess of 90% being removed, and there is no significant difference if 8-30 ethylene oxide groups are present [33].

The fluorescent whitening agents which are added to detergent formulations are removed from the effluent during wastewater treatment, mainly as part of the sludge [34]. The material is not readily translocated by rainwater and so is not taken up by plants.

The presence of phosphate in detergents can lead to problems of eutrophication, and a ban on the use of this can lead to a considerable saving in costs for phosphate removal from effluents [35]. Trisodium nitrilotriacetate (NTA) is being considered as a partial replacement for phosphate builders, because it is biodegradable. The NTA in a domestic wastewater system is degraded at concentrations up to 30 mg/l and the process does not

Fig.10

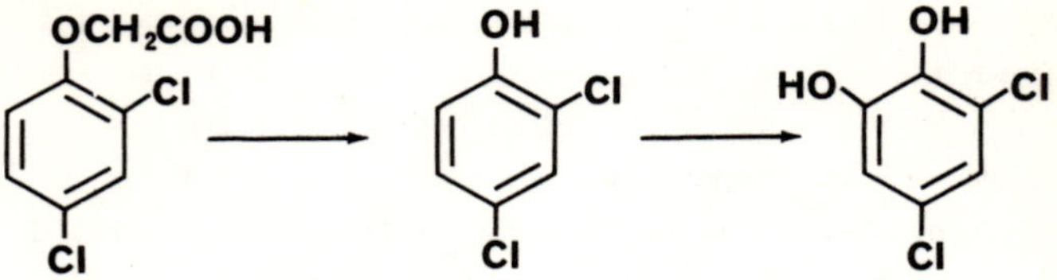

2,4-D

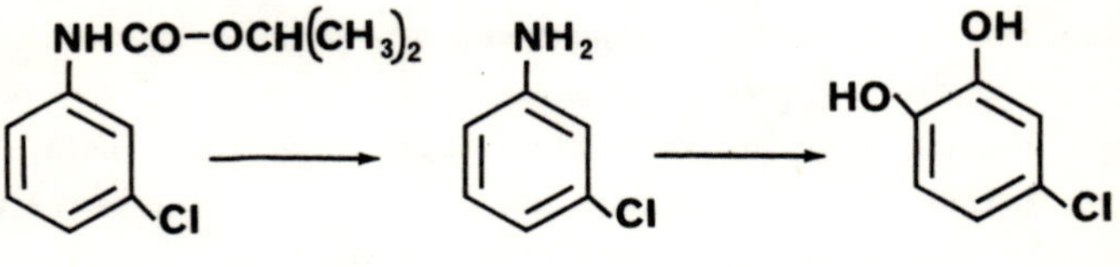

CIPC

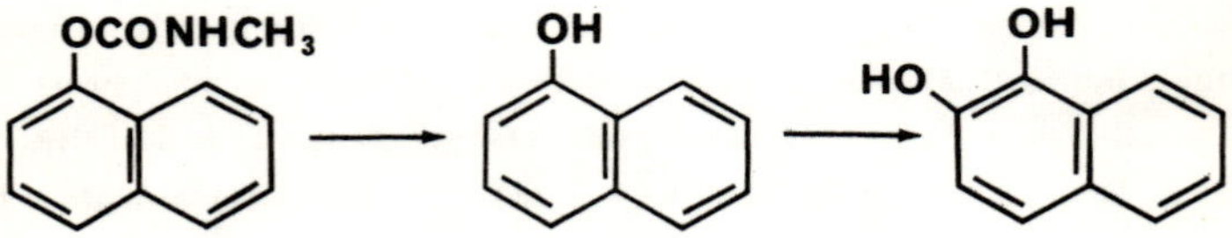

Carbaryl

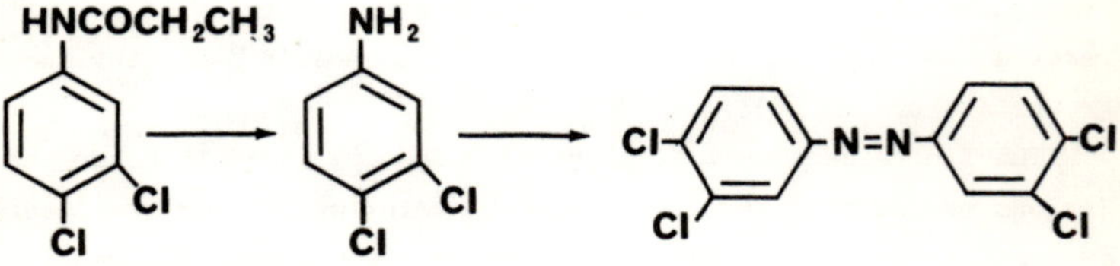

Propanil

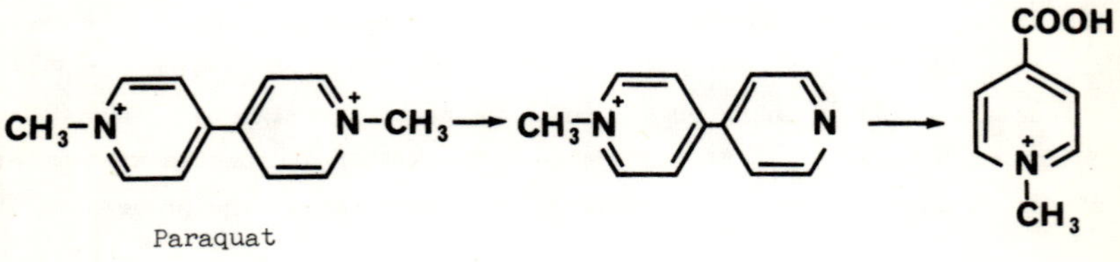

Paraquat

appear to be influenced by cold weather [36]. If a 5-7 wk acclimatisation
is allowed, then NTA can be removed in oxidation ponds [37]. NTA does not
give any abnormal corrosion problems, nor is it toxic to fish at a 1:15
dilution and so could be a valid replacement for some of the phosphate.

Pesticides

Many herbicides and pesticides such as the phenyl carbamates, phenyl
ureas and acylanilides can be degraded as shown in Figs. 2 and 3, although
pesticides such as propanil which give rise to 3:4-dichloroaniline are
persistant.

The initial reactions in the degradation of Diuron and 2:4-dichloro-
phenoxyacetic acid (2;4D) are shown in Figs. 1(b) and 1(c). The presence
of such compounds in the environment will obviously favour micro-organisms
capable of degrading them and a number of species [38,39,40,41] degrading
pesticides and herbicides have been reported. The types of reaction found
in degradation are shown in Fig. 10.

The presence or absence of some species has been used as a bioassay
for certain compounds as the deleterious effect of organo-halides on
certain species is fairly well documented [42,43,44,45,46]. The synthesis
of soft pesticides has been discussed by Kaufman and Plimmer [47] and the
use of third generation pesticides is considered by Williams [48].

However the biodegradation of pesticides which is usually considered
to be a desirable property can cause problems. The herbicide propanil is
partially degraded to form 3:4-dichloroaniline which can form azo-compounds
of a type known to cause cancer.

It can be seen from Fig. 10 that many herbicides and pesticides contain one or more halogen atoms present in the aromatic ring. The presence
of these atoms frequently reduces the biodegradability of the molecule,
thus increasing its persistance [49]. In addition to this reduced susceptibility to oxidation, if partial oxidation takes place then the substituted metabolite is frequently resistant to further degradation by the
more common enzymes [50]. Substitution in the meta position usually
produces a molecule considerably more resistant than substitutions in other
positions and increasing the number of halogen atoms also increases resistance [51]. The influence of methyl and alkyl substituents on aromatic
degradation is discussed by Treccani [52].

Cometabolism is an important method for partial degradation of many
organo halides. For example, 3-chlorocatechol is derived from chloro-

benzoic acid by bacteria utilizing benzoic acid for growth and similar phenomena have been reported [53,54,55] by other authors.

Removal of the halogen group may take place by means of several routes. Reductive dehalogenation [56] can take place in an anaerobic environment, and hydrolysis, and dehydrodehalogenation to form an olefin have also been shown.

There does exist a distinct possibility that halogenated compounds may be degraded by enzymes not yet characterized. The list of naturally occurring halogenated compounds is increasing very rapidly [57], especially those of marine origin, and it is probable that a number of enzymes cleaving carbon-halogen bonds remain to be discovered.

Research extremely relevant to this topic has recently been published [58]. The degradation of the herbicide Dalapon (2:2-dichloropropionic acid) was studied in a chemostat for approximately 3,700 generations (13,500 h). The stable microbial community utilizing Dalapon as the sole carbon source consisted of six organisms, three of which were capable of growing upon Dalapon as the sole carbon source. These were referred to as the primary utilizers. The remaining three organisms (secondary utilizers), one of which was Pseudomonas putida, could not grow upon Dalapon and in the chemo-stat were presumably using metabolites or excretory matter produced by the primary utilizers.

After approximately 800 generations, a fourth primary utilizer appeared which was identified as Ps. putida and appeared identical with the secondary utilizer of that species. Biochemical examination of the parent and mutant strains, and growth as pure cultures for long periods in a chemostat, showed that the mutant had acquired an enzyme called dehalogenase capable of hydrolysing Dalapon to pyruvic acid and free chloride. Examination of the parent and mutant in pure culture showed that the enzyme had been acquired by a mutational event, and not by genetic acquisition from other organisms.

One of the most interesting points in this paper is the stable relationships of the community and the fact that a fourth primary utilizer could evolve after such a long period. In the past studies on mixed bacterial populations in chemostats have tended to suggest that the most competitive of the organisms would eventually become the dominant species, excluding all the other species. This work appears to contradict that theory, and the authors indicate that a number of as yet unidentified interactions are occurring which tend to stabilize the community.

Another very important point made is that a considerable amount of
time was needed to allow the mutant to become apparent. During this
period of time the organisms obviously required an energy source which was
the metabolite excreted by the primary producers. This raises a strong
suspicion that molecules previously called recalcitrant may have acquired
this label after being tested for an insufficient period of time in the
absence of an alternative carbon source.

It is to be hoped that this work stimulates chemostat studies of other
more complex pesticides, although it should be stressed that Dalapon
represents a relatively simple biochemical problem when compared with a
compound such as Mirex which has the formula $C_{10}Cl_{12}$.

As a final note, several authors [59,60] have suggested treating the
ecosystem with micro-organisms especially adapted to some pollutants in an
effort to dissipate the pollutant more rapidly [61].

REFERENCES

1   S. Dagley, in P.N. Campbell and W.N. Aldridge (Editors), Essays in
    Biochemistry, Vol. 11, Academic Press, 1975, p. 81.
2   M. Blumer, Scientific American, 234(1976)35.
3   P.H. Clarke and L.N. Ornston, in P.H. Clarke and M.H. Richmond
    (Editors), Genetics and Biochemistry of Pseudomonas, Wiley-Interscience,
    1975, Ch. 7, p. 191.
4   P.H. Clarke and L.N. Ornston, ibid. Ch. 8, p. 263.
5   K. Murray and P.A. Williams, J. Bact., 117(1974)1153.
6   J.I. Davies and W.C. Evans, Biochem. J., 91(1964)251.
7   L.N. Ornston, Bact. Revs., 35(1971)87.
8   S. Dagley, Advances in Microbial Physiology, 6(1971)1.
9   J. Mandelstam, in J. Mandelstam and K. McQuillen (Editors),
    Biochemistry of Bacterial Growth, Blackwell, Oxford, 1968, p. 414.
10 L.N. Ornston and D. Parker, Biochem. Soc. Trans., 4(1976)468.
11 A.L. Lehninger, Biochemistry, Worth Publishers, New York, 2nd ed.,
    1975, Ch. 35, p. 977.
12 G.D. Hegeman, J. Bact., 91(1966)1140.
13 J. Mandelstam and G.A. Jacoby, Biochem. J., 94(1965)569.
14 D. Parker and L.N. Ornston, J. Bact., 126(1976)272.

214

15 D.T. Gibson, in J.M. Sharpley and A.M. Kaplan (Editors), Proc. 3rd Int.
    Biodegradation Symposium, Applied Science Publishers, 1976, p. 57.

16 P.J. Chapman and D.W. Ribbons, J. Bact., 125(1976)975.

17 A.M. Chakrabarty, J. Bact., 112(1972)815.

18 P.A. Williams and K. Murray, J. Bact., 120(1974)416.

19 M.J. Worsey and P.A. Williams, J. Bact., 124(1975)7.

20 P.A. Williams and M.J. Worsey, J. Bact., 125(1976)818.

21 N.W. Dunn and I.C. Gunsalus, J. Bact., 114(1973)974.

22 S. Benson and J. Shapiro, J. Bact., 126(1976)794.

23 M.L. Whellis, Advances in Microbiology, 29(1975)505.

24 H.R. Mahler and E.H. Cordes, Biological Chemistry, Harper & Row, New
    York, 2nd ed., 1971, Ch. 18, p. 819.

25 A.G. Callely, in B. Spencer (Editor), Industrial Aspects of Biochemistry
    Fed. European Biochem. Soc., 1974, p. 515.

26 P.W. Trudgill, Biochem. Soc. Trans., 4(1976)458.

27 N.A. Donaghue, M. Griffin, D.B. Norris and P.W. Trudgill, in
    J.M. Sharpley and A.M. Kaplan (Editors), Proc. 3rd Int. Biodegradation
    Symposium, Applied Science Publishers, 1976, p. 43.

28 W.J. Payne, W.J. Wiebe and R.R. Christian, Bioscience, 20(1970)862.

29 G.V. Stennett and G.E. Eden, Wat. Res., 5(1971)601.

30 W.E. Gledhill, Appl. Microbiol., 30(1975)922.

31 O.E.C.D. Pollution by Detergents. Determination of the Biodegradability
    of Anionic Synthetic Surface Active Agents, Paris, 1971.

32 W. Janicke and W. Niemitz, Vom Wasser, 40(1973)369.

33 L. Rudling and P. Solyom, Water Res., 8(1974)115.

34 C.R. Ganz, C. Liebert, J. Schulze and P.S. Stensby, J. Water Pollut.
    Control Fed., 47(1975)2834.

35 P. Pieczonka and N.E. Hopson, Wat. Sewage Works, 121(1974)52.

36 C.E. Renn, J. Water Pollut. Control Fed., 46(1974)2363.

37 S. Klein, J. Water Pollut. Control Fed., 46(1974)78.

38 S.J.L. Wright, in G. Sykes and F.A. Skinner (Editors), Microbial
    Aspects of Pollution, Academic Press, 1971, p. 233.

39 S.J.L. Wright, in B. Spencer (Editor), Industrial Aspects of
    Biochemistry, Federation of European Biochemical Societies, 1974, p.495.

40 I.J. Higgins and R.G. Burns, The Chemistry and Microbiology of
    Pollution, Academic Press, 1975, Ch. 2, p. 7.

41 C.S. Helling, J. Environ. Quality, 5(1976)1.

42 D.A. Ratcliffe, J. Appl. Ecol., 7(1970)67.

43 J.R. Robinson, Bird Study, 17(1970)195.

44 D.S. Miller, W.B. Kinter and D.B. Peakall, Nature, 259(1976)122.

45 P.W. Trudgill and S.R.H. Bowering, J. Gen. Microbiol., 73(1972)577.

46 I.H. Suffet, S. Friant, C. Marcinkiewicz, J. McGuire and D.T-L. Wong, J. Water Poll. Control Fed., 47(1975)1169.

47 D.D. Kaufman and J.R. Plimmer, in R. Mitchell (Editor), Water Pollution Microbiology, Wiley-Interscience, 1972, p. 173.

48 C.M. Williams, Scientific American, 217(1967)13.

49 S. Dagley, Biochem. Soc. Trans., 4(1976)455.

50 P. Chapman, Biochem. Soc. Trans., 4(1976)463.

51 D.D. Kaufmann, J. Agr. Food Chem., 15(1967)582.

52 V. Trecanni, in B. Spencer (Editor), Industrial Aspects of Biochemistry, Federation of European Biochemical Societies, 1974, p. 533.

53 N. Walker, Soil Biol. Biochem., 5(1973)525.

54 M. Ahmed and D.D. Focht, Canad. J. Microbiol., 19(1973)47.

55 P. Goldman, Degradation of Synthetic Molecules in the Biosphere, Nat. Acad. Sci. U.S., Washington D.C., 1972, p. 147.

56 P.C. Kearney, D.D. Kaufmann and M.L. Beall, Biochem. Biophys. Res. Comm., 14(1964)29.

57 J.F. Siuda and J.F. De Bernadis, Lloydia, 36(1973)107.

58 E. Senior, A.T. Bull and J.H. Slater, Nature, 263(1976)476.

59 C.G. Clark and S.J.L. Wright, Soil Biol. Biochem., 2(1970)19.

60 G.W. McClure, J. Environ. Qual., 1(1972)177.

61 D.A.F. Rillo, J.R. Mylroie and A.M. Chakrabarty, in J.M. Sharpley and A.M. Kaplan (Editors), Proc. 3rd Int. Biodegradation Symposium, Applied Science Publishers, 1976, p. 205.